# BASICS OF NUCLEAR PHYSICS

# BASICS OF NUCLEAR PHYSICS

RAJNISH PANDIT

**ANMOL PUBLICATIONS PVT. LTD.**
NEW DELHI - 110 002 (INDIA)

**ANMOL PUBLICATIONS PVT. LTD.**
**H.O.:** 4374/4B, Ansari Road, Darya Ganj,
New Delhi-110 002 (India)
Ph.: 23278000, 23261597
**B.O.:** No. 1015, Ist Main Road, BSK IIIrd Stage
IIIrd Phase, IIIrd Block,
Bangalore - 560 085 (India)
*Visit us at*: www.anmolpublications.com

*Basics of Nuclear Physics*



ISBN 978-81-261-3588-2

PRINTED IN INDIA

Printed at Mehra Offset Press, Delhi.

# Contents

# Preface

Nuclear Physics is the study of the components, structure and behaviour of the nucleus of the atom. It is especially concerned with the nature of matter and with nuclear energy. Nuclear physicists investigate nuclear matter on all scales, from sub-atomic particles to supernovas. Research areas include the structure of the nucleus at different temperatures and pressures, the origin of elements and the structure and evolution of stars.

The contribution of nuclear physics to energy sector, medical sciences, biology, radiobiology and to many other fields is now well recognised. The widespread applications of nuclear techniques and their strong impact on society require an education of the general public in the concept of elementary nuclear phenomena.

Basics of Nuclear Physics is written from an experimental point of view, it discusses the fundamental concepts and practical applications of nuclear physics in detail. The book will be highly useful for students, teachers and all those concerned.

Author

## Chapter 1

# Introduction to Nuclear Physics

### NUCLEAR PHYSICS

The First World War lasted from 1914 to 1918. Geiger and Marsden were both at the Western front, on opposite sides. Rutherford had a large water tank installed on the ground floor of the building in Manchester, to carry out research on defence against submarine attack. Nevertheless, occasional research on alpha scattering continued. Scattering from heavy nuclei was fully accounted for by the electrostatic repulsion, so Rutherford concentrated on light nuclei, including hydrogen and nitrogen. In 1919, Rutherford established that an alpha impinging on a nitrogen nucleus can cause a *hydrogen* atom to appear! Newspaper headlines blared that Rutherford had "split the atom".

Shortly after that experiment, Rutherford moved back to Cambridge to succeed J. J. Thomson as head of the Cavendish laboratory, working with one of his former students, Chadwick, who had spent the war years interned in Germany. They discovered many unusual effects with alpha scattering from light nuclei. In 1921, Chadwick and co-author Bieler wrote: "The present experiments do not seem to throw any light on the nature of the law of variation of the forces at the seat of an electric charge, but merely show that the forces are of great intensity ... It is our task to find some field of force which will reproduce these effects."

In fact, Rutherford was beginning to focus his attention on the actual construction of the nucleus and the alpha particle.

He coined the word "proton" to describe the hydrogen nucleus, it first appeared in print in 1920. At first, he thought the alpha must be made up of four of these protons somehow bound together by having two electrons in the middle - this would get the mass and charge right, but of course nobody could construct a plausible electrostatic configuration. Then he had the idea that maybe there was a special very tightly bound state of a proton and an electron, much smaller than an atom. By 1924, he and Chadwick were discussing how to detect this neutron. It wasn't going to be easy - it probably wouldn't leave much of a track in a cloud chamber. In fact, Chadwick did discover the neutron, but not until 1932, and it wasn't much like their imagined proton-electron bound state. But it did usher in the modern era in nuclear physics.

The relevant length scale for measuring nuclear size is the femtometer 1 fm = $10^{-15}$ meters. Physicists usually call this length one *fermi*. Nuclei vary from about one to a few fermis in radius. Recall that the Bohr radius of Hydrogen is of order $10^{-10}$ meters, so the nucleus is far smaller than the atom. Nuclear size was first measured by Rutherford, by noting how close ±-particles came to the nucleus before the scattering ceased to be pure Coulomb repulsion (at which point they were actually hitting the nuclear surface). Despite its small size, the nucleus has about 99.97per cent of the mass of the atom (a bit less for hydrogen).

The simplest nucleus, that of hydrogen, is a single proton: an elementary particle of mass about 940 MeV, carrying positive charge exactly opposite to the electron's charge, having a spin of one half and being a fermion (so no two protons can be in the same quantum state).

The next simplest nucleus, called the *deuteron*, is a bound state of a proton and a neutron. The neutron, like the proton, is a spin one-half fermion, but it has no electric charge, and is slightly heavier (by 1.3 MeV) than the proton. The binding energy of the deuteron (analogous to the 13.6 eV for the Hydrogen atom) is 2.2MeV. A photon of this energy could "ionize" the deuteron into a separated proton and neutron.

However, it is not necessary to actually do this experiment to establish how tightly the deuteron is bound. One need only weigh the deuteron accurately. It has a mass of 1875.61MeV. The proton has a mass of 938.27MeV, the neutron 939.56MeV, so together (but some distance apart!) they have a mass of 1877.93MeV, 2.2MeV *more* than the deuteron. Thus, when a proton and a neutron come together to form a deuteron, they must unload 2.2MeV of energy, which they do by emitting a photon.

Both protons and neutrons, being fermions, obey the exclusion principle, two protons with spin up cannot be in the same state, although two with opposite spin directions could, and a proton and a neutron can occupy the same spot at the same time!

Protons and neutrons are referred to as *nucleons*. The total number of nucleons in a nucleus is usually denoted by $A$, where $A = Z + N$, $Z$ protons and $N$ neutrons. The chemical properties of an atom are determined by the number of electrons, the same as the number of protons $Z$. This is called the *atomic number*. Nuclei can have the same atomic number, but different numbers of neutrons. These nuclei are called *isotopes*, the Greek for "same place", since they are in the same place in the periodic table.

These nucleons attract each other with a short range but very strong force, called the *nuclear force*. The situation here is different from that for electrons in the atom, where the strong central force tends to dominate. In the nucleus the nucleons are attracted mainly by their immediate neighbours. Nevertheless, it is a useful beginning to think of this attractive force as being a potential well, as seen by an individual nucleon, and think in terms of the nuclei as filling the lowest available quantum states in this well, just as we did for electrons in the atom. We find that the Helium nucleus, $2p+2n$, (a.k.a. the ±-particle) is tightly bound—the four nucleons can all occupy the lowest state in the well. However, some larger nuclei, like C, O, Fe are actually a little more tightly bound even than He (about 8.5 MeV per nucleon as opposed to about

7.5 for He) because each nucleon is attracted to its close partners, and there are more close partners in these larger nuclei. It has also been argued that some of these higher nuclei strongly resemble bound states of ±-particles. The total binding energy (usually expressed per nucleon) of any nucleus is easy to find—just as for the deuteron above, the mass of the nucleus is found accurately, and subtracted from the sum of the masses of the separate nucleons.

## COULOMB REPULSION AND α DECAY

As one goes to really large nuclei, the binding per nucleon actually decreases. This is a consequence of the electrostatic repulsion between the protons. The essential point is that the Coulomb repulsion is *long range*, but the nuclear attraction is short range. Considering a proton at the surface of a nucleus, it is bound to the nucleus by the attractive forces of its nearest neighbours, say five or six of them, and this number is little affected as one goes up the periodic table from, say, Fe to U. On the other hand, the proton on the surface is being pushed away by the electrostatic repulsion from the other protons, and this increases substantially as the number of *other* protons goes from 25 in iron to 91 in uranium, even though they may be a little further away on average. In fact, this is why there is a limit on the size of atoms.

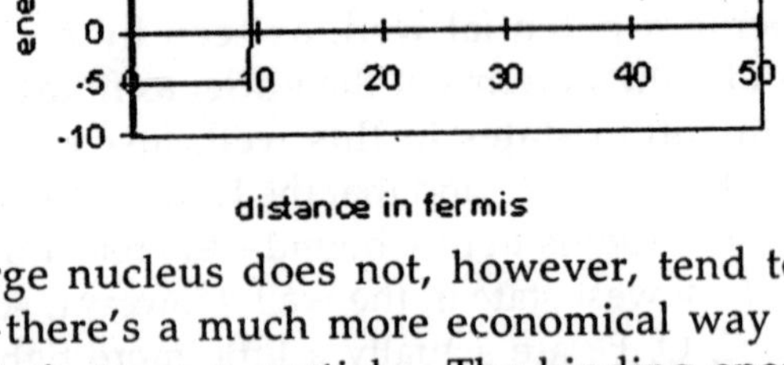


A large nucleus does not, however, tend to spit out protons—there's a much more economical way to unload charge. It ejects an ±-particle. The binding energy of the nucleons in an ± is almost exactly the same as it is in a U nucleus, so little nuclear binding has to be traded for lowering the electrostatic potential energy.

Thus the most common form of Uranium, $U^{238}$ ($A$=238, $Z$=92) goes to Thorium ($A$=234, $Z$=90) by ±-decay. The surprising thing is, it takes a time comparable to the age of the universe to do so! If it's energetically desirable, why does it take so long?

The answer is—there's a barrier in the way. It is a good model to imagine the U to have the a particle inside it, confined to a deep square well potential representing the attractive nuclear force. However, the α-particle wavefunction will penetrate some distance into the wall of this finite square well potential, where the α will be away from the nuclear force, but not away from the long-range electrostatic repulsion of the other protons in the nucleus. The picture is, then, that outside the well, the ground is not flat—it's like a well at the top of a hill. The implication of this is that if the energy level of the a inside the well is higher than the flat land surrounding the hill, the α-wavefunction after dying away underground as it tunnels out from the well will re-emerge at some distance, oscillating. In other words, there is some finite probability of finding the α shooting away from the nucleus some distance away. The actual probability of this happening depends on how much the wavefunction decayed as it tunneled underground away from the well, which depended on how far its energy was below the ground. For $U^{238}$, this decay on tunneling away from the well is so rapid that the α takes on average billions of years to escape. However, many different nuclei undergo α decay, the quickest in millionths of a second. The energies of the emitted α's can be measured, and confirm this explanation of quantum mechanical tunneling, first put forward in 1928. The quickest decays correspond to the most energetic ±'s, just as the tunneling picture would suggest.

## β-DECAY

These heavier nuclei have more neutrons than protons, because the Coulomb repulsion makes it harder to bind protons. However, a price is paid—thinking of a shell-type model, the extra neutrons go into higher energy shells. Thus for nuclei of a given size, there is an optimum ratio of neutrons to protons, close to one for small nuclei where the Coulomb effects are small enough to be neglected, close to 1.5 for the largest nuclei.

Note that if a large nucleus decays by emitting an ±, the nucleus left behind has an even greater ratio of neutrons to protons, since $\alpha = 2p+2n$. This implies that it might be able to go to a lower energy state if it could trade a neutron for a proton. In fact, this can sometimes be done. The neutron itself is not a stable particle. A neutron alone in space will only last about ten minutes before ejecting an electron, leaving a proton behind. This is energetically possible, because the mass of the neutron exceeds the sum of the masses of the proton and electron. The process is called β-decay. It can also occur in a nucleus, but now the different binding energies of the initial neutron and the final proton must be factored in to decide if it will go. Other things being equal, the more neutrons there are present in the nucleus, the more likely it is to β-decay. Actually, β-decay was a big puzzle when it was first carefully observed, because it was discovered that identical nuclei β-decaying to the same final states ejected electrons over a range of energies, up to a maximum corresponding to the energy difference between the initial and final nuclei. This led to the suggestion that maybe, after all, energy wasn't conserved in every process in nature.

But physicists are a conservative lot when it comes to energy conservation, and Pauli suggested that maybe $n \rightarrow p+e$ wasn't the whole story—some other particle was being created at the same time, carrying away the balance of the energy, but *undetected*. This meant it had no charge, no mass, and, unlike the photon, no electric interaction. Fermi called it the *neutrino*, the little neutral one. Another reason the neutrino seemed necessary was that the initial neutron had spin one-half, as did the final proton *and* the emitted electron. It was difficult to see how angular momentum could be conserved if no other particle were emitted, because even if the final proton and electron had some orbital angular momentum, that comes in whole-number amounts, so how could two half-integer spins and an integer angular momentum add to be equal to the initial half-integer spin? It turned out they couldn't, the neutrino really was there all along, it had spin one-half itself, but it was years before it was detected. Detecting neutrinos is still a difficult business, since they are not sensitive to the nuclear (strong) force, nor to electromagnetic fields. The force that causes neutron (and other particle) decay is called the weak force. The only way to detect a neutrino is if it collides with a particle and interacts weakly essentially reversing one of these decays. The problem is that a neutrino moving downwards is extremely unlikely to interact with a nucleus, it will almost certainly just pass through the earth. Therefore, to detect neutrinos takes large detectors and a strong flux of neutrinos. Much effort has been expended detecting neutrinos from the sun, since they are our only window into the nuclear processes taking place deep in the sun—they can get out unscathed, unlike photons or any other particle emitted. In fact, it is found that the number detected is less by a factor of two or three from that predicted using standard analyses of the solar nuclear processes. It turns out that in fact the neutrino is not quite massless, and there are actually three kinds of neutrinos! Furthermore, as a neutrino passes through matter, there is a nonzero probability of it changing from one kind to another. The solar nuclear processes generate what are called electron neutrinos, and the initial efforts at neutrino detection looked for that kind. More recent detectors count the other two kinds (mu neutrinos and tau neutrinos) as well, and the total number of neutrinos detected agrees well with the prediction of the total number generated in the sun, so it is now believed that some of the original electron neutrinos change types traveling from their point of origin to the sun's surface.

Anyway, back to the radioactive decay of $U^{238}$. It emits an $\alpha$ to become $Th^{234}$, then $\beta$ decays twice in succession to become $U^{234}$. This is now relatively proton-rich, and in fact a decays five times in succession to become a form of lead,

$U^{234} \rightarrow Th^{230} \rightarrow Ra^{226} \rightarrow Rn^{222} \rightarrow Po^{218} \rightarrow Pb^{214}$

$Pb^{214}$ undergoes two ² decays, an ± decay and two more β decays before it reaches stability at $Pb^{206}$. The sequence above is through metallic elements with one exception—Rn, radon, is a gas because, atomically speaking, it has completely filled electron shells, and so is chemically unreactive, and gaseous at ordinary temperatures. It emits its ± particle in a period of days, which gives it time to diffuse out of the rock into somebody's basement.

## γ-DECAY

The three forms of radiation emitted by decaying nuclei, α, β and γ, were named before the particles being emitted were fully identified. It was only realised later that the β-rays were electrons. The ³ rays are just high-energy photons, of order 100 keV to a few MeV. Emission of γ rays is similar (but at much higher energies) to emission of photons by excited states of atoms. The nucleus can be excited by having just emitted an α or β, or by colliding with another nucleus, or being bombarded by neutrons, say. All these events can lead to a nucleus in which the charge distribution is oscillating, and electromagnetic radiation ensues.

## THE DROPLET MODEL

In some ways, the binding of nucleons in a nucleus is reminiscent of that of molecules in a drop of water—the force is attractive between near neighbours. Notice how different this is from the electrons in an atom, held in place by a large central attracting object. This "droplet" model of the nucleus is useful for understanding nuclear fission. If the droplet is somehow caused to vibrate vigorously, it might break into two, or more, smaller nuclei. For the $U^{235}$ nucleus (only 0.7per cent in naturally occurring uranium), if a slow neutron gets too close, the attractive nuclear force pulls it towards the nucleus so strongly that in the resulting collision, the nucleus breaks into two smaller ones, and two or three neutrons are ejected (since the smaller nuclei formed have lower proportions of neutrons). There is also substantial energy release—about 200

MeV. The much more common $U^{238}$ nucleus also attracts free neutrons, but not so strongly, and it survives neutron capture without fission.

A neutron has to be moving very slowly to be captured by $U^{235}$. On the other hand, when $U^{235}$ fissions the emitted neutrons are moving very fast. To get a *chain reaction* in a nuclear reactor, the main problem is to slow down the emitted neutrons before they leave the reactor. If they can be slowed down successfully, neutrons emitted by one fission will initiate further fissions, giving continuous—perhaps increasing—energy output. However, the neutrons bounce off the heavy uranium nuclei with almost no energy loss—to get a particle to lose energy in a collision, it must collide with something approximately its own size. Water is often used, the neutrons lose energy bouncing off the protons in the hydrogen atoms. Unfortunately, the neutrons often stick to the protons, forming deuterons, so to keep the reaction going the uranium must be enriched—the $U^{235}$ content boosted to one to four percent. Of course, care must be taken to ensure that the chain reaction isn't *too* successful—moderators, which absorb neutrons, are put into the reactor if the rate of fission increases beyond the planned level. Needless to say, moderators are not used in military applications of this concept.

## NUCLEAR PHYSICS

### FISSION AND FUSION

Fission of radioactive elements was already well established in the early part of the century, and activation by neutrons, to generate more unstable isotopes, was investigated before fission of natural isotopes was seen. The inverse process, fusion, was understood somewhat later, and Niels Bohr developped a model describing the nucleus as a fluid drop. This model - the collective model - was further developped by his son Aage Bohr and Ben Mottelson. A very different model of the nucleus, the shell model, was designed by Maria Goeppert-Mayer and Hans Jensen in 1952, concentrating on individual nucleons. The dichotomy between a description as individual particles and as a collective whole characterises much of "low-energy" nuclear physics.

## LOW-ENERGY NUCLEAR PHYSICS

The field of low-energy nuclear physics, which concentrates mainly on structure of and low-energy reaction on nuclei, has become one of the smaller parts of nuclear physics (apart from in the UK). Notable results have included better understanding of the nuclear medium, high-spin physics, superdeformation and halo nuclei. Current experimental interest is in those nuclei near the "driplines" which are of astrophysical importance, as well as of other interest.

## MEDIUM-ENERGY NUCLEAR PHYSICS

Medium energy nuclear physics is interested in the response of a nucleus to probes at such energies that we can no longer consider nucleons to be elementary particles. Most modern experiments are done by electron scattering, and concentrate on the role of QCD in nuclei, the structure of mesons in nuclei and other complicated questions.

## HIGH-ENERGY NUCLEAR PHYSICS

This is not a very well-defined field, since particle physicists are also working here. It is mainly concerned with ultra-relativistic scattering of nuclei from each other, addressing questions about the quark-gluon plasma. It should be nuclear physics, since we consider "dirty" systems of many particles, which are what nuclear physicists are good at.

## MESONS, LEPTONS AND NEUTRINOS

In 1934 Yukawa introduces a new particle, the pion ($\pi$), which can be used to describe nuclear binding. He estimates it's mass at 200 electron masses. In 1937 such a particle is first seen in cosmic rays. It is later realised that it interacts too weakly to be the pion and is actually a lepton (electron-like particle) called the $\mu$. The is found (in cosmic rays) and is the progenitor of the $\mu$'s that were seen before:

$$\pi^+ \rightarrow \mu^+ + \nu_\mu$$

The next year artificial pions are produced in an accelerator, and in 1950 the neutral pion is found,

$$\pi^0 \rightarrow \gamma\gamma .$$

This is an example of the conservation of electric charge. Already in 1938 Stuckelberg had found that there are other conserved quantities: the number of baryons ($n$ and $p$ and...) is also conserved!

After a serious break in the work during the latter part of WWII, activity resumed again. The theory of electrons and positrons interacting through the electromagnetic field (photons) was tackled seriously, and with important contributions of (amongst others) Tomonaga, Schwinger and Feynman was developed into a highly accurate tool to describe hyperfine structure.

Experimental activity also resumed. Cosmic rays still provided an important source of extremely energetic particles, and in 1947 a "strange" particle ($K^+$ was discovered through its very peculiar decay pattern. Balloon experiments led to additional discoveries: So-called V particles were found, which were neutral particles, identified as the $\Lambda^0$ and $K^0$. It was realised that a new conserved quantity had been found. It was called strangeness.

The technological development around WWII led to an explosion in the use of accelerators, and more and more particles were found. A few of the important ones are the antiproton, which was first seen in 1955, and the $\Delta$, a very peculiar excited state of the nucleon, that comes in four charge states $\Delta^{++}, \Delta^+, \Delta^0, \Delta^-$.

Theory was develop-ping rapidly as well. A few highlights: In 1954 Yang and Mills develop the concept of gauged Yang-Mills fields. It looked like a mathematical game at the time, but it proved to be the key tool in developing what is now called "the standard model".

In 1956 Yang and Lee make the revolutionary suggestion that parity is not necessarily conserved in the weak interactions. In the same year "madam" CS Wu and Alder show experimentally that this is true: God is weakly left-handed!

In 1957 Schwinger, Bludman and Glashow suggest that all weak interactions (radioactive decay) are mediated by the charged bosons $W^{\pm}$. In 1961 Gell-Mann and Ne'eman introduce the "eightfold way": a mathematical taxonomy to organize the particle zoo.

## THE SUB-STRUCTURE OF THE NUCLEON (QCD)

In 1964 Gell-mann and Zweig introduce the idea of quarks: particles with spin 1/2 and fractional charges. They are called up, down and strange and have charges 2/3, - 1/3, - 1/3 times the electron charge.

Since it was found (in 1962) that electrons and muons are each accompanied by their own neutrino, it is proposed to organize the quarks in multiplets as well:

$e\upsilon_e\,(u, d\,)$

$\mu\nu_\mu\,(s, c)$

This requires a fourth quark, which is called charm.

In 1965 Greenberg, Han and Nambu explain why we can't see quarks: quarks carry colour charge, and all observe particles must have colour charge 0. Mesons have a quark and an antiquark, and baryons must be build from three quarks through its peculiar symmetry.

The first evidence of quarks is found (1969) in an experiment at SLAC, where small pips inside the proton are seen. This gives additional impetus to develop a theory that incorporates some of the ideas already found: this is called QCD. It is shown that even though quarks and gluons (the building blocks of the theory) exist, they cannot be created as free particles. At very high energies (very short distances) it is found that they behave more and more like real free particles.

This explains the SLAC experiment, and is called asymptotic freedom.

The $J/\psi$ meson is discovered in 1974, and proves to be the $c\bar{c}$ bound state. Other mesons are discovered (D0, $\overline{\mu}$ $c$) and agree with QCD.

In 1976 a third lepton, a heavy electron, is discovered ($\tau$). This was unexpected! A matching quark (b for bottom or beauty) is found in 1977. Where is its partner, the top? It will only be found in 1995, and has a mass of 175 GeV/$c^2$ Together with the conclusion that there are no further light neutrinos (and one might hope no quarks and charged leptons) this closes a chapter in particle physics.

## THE $W^{\pm}$AND Z BOSONS

On the other side a electro-weak interaction is developed by Weinberg and Salam. A few years later 't Hooft shows that it is a well-posed theory. This predicts the existence of three extremely heavy bosons that mediate the weak force: the $Z^0$ and the $W^{\pm}$. These have been found in 1983. There is one more particle predicted by these theories: the Higgs particle. Must be very heavy!

## GUTS, SUPERSYMMETRY, SUPERGRAVITY

This is not the end of the story. The standard model is surprisingly inelegant, and contains way to many parameters for theorists to be happy. There is a dark mass problem in astrophysics - most of the mass in the universe is not seen! This all leads to the idea of an underlying theory. Many different ideas have been developed, but experiment will have the last word! It might already be getting some signals: researchers at DESY see a new signal in a region of particle that are 200 GeV heavy - it might be noise, but it could well be significant!

There are several ideas floating around: one is the grand-unified theory, where we try to comine all the disparate forces in nature in one big theoretical frame. Not unrelated is the idea of supersymmetries: For every "boson" we have a "fermion". There are some indications that such theories may actually be able to make useful predictions.

## EXTRATERRESTRIAL PARTICLE PHYSICS

One of the problems is that it is difficult to see how we can actually build a microscope that can look a a small enough scale, i.e., how we can build an accelerator that will be able to accelarte particles to high enough energies? The answer is simple - and has been more or less the same through the years: Look at the cosmos. Processes on an astrophysical scale can have amazing energies.

## BALLOON EXPERIMENTS

One of the most used techniques is to use balloons to send up some instrumentation. Once the atmosphere is no longer the perturbing factor it normally is, one can then try to detect interesting physics. A problem is the relatively limited payload that can be carried by a balloon.

## GROUND BASED SYSTEMS

These days people concentrate on those rare, extremely high energy processes (of about $10^{29}$ eV), where the effect of the atmosphere actually help detection. The trick is to look at showers of (lower-energy) particles created when such a high-energy particle travels through the earth's atmosphere.

## DARK MATTER

One of the interesting cosmological questions is whether we live in an open or closed universe. From various measurements we seem to get conflicting indications about the mass density of (parts of) the universe. It seems that the ration of luminous to non-luminous matter is rather small. Where is all that "dark mass": Mini-jupiters, small planetoids, dust, or new particles.

## (SOLAR) NEUTRINOS

The neutrino is a very interesting particle. Even though we believe that we understand the nuclear physics of the sun, the number of neutrinos emitted from the sun seems to anomalously small. Unfortunately this is very hard to measure, and one needs quite a few different experiments to disentangle the physics behind these processes. Such experiments are coming on line in the next few years. These can also look at neutrinos coming from other astrophysical sources, such as supernovas, and enhance our understanding of those processes. Current indications from Kamiokande are that neutrinos do have mass, but oscillation problems still need to be resolved.

## Chapter 2

# The Atomic Nucleus

### STRUCTURE OF MATTER

Early Greek philosophers speculated that the earth was made up of different combinations of basic substances, or elements. They considered these basic elements to be earth, air, water, and fire. Modern science shows that the early Greeks held the correct concept that matter consists of a combination of basic elements, but they incorrectly identified the elements.

In 1661 the English chemist Robert Boyle published the modern criterion for an element. He defined an element to be a basic substance that cannot be broken down into any simpler substance after it is isolated from a compound, but can be combined with other elements to form compounds. To date, 105 different elements have been confirmed to exist, and researchers claim to have discovered three additional elements. Of the 105 confirmed elements, 90 exist in nature and 15 are man-made.

Another basic concept of matter that the Greeks debated was whether matter was continuous or discrete. That is, whether matter could be continuously divided and subdivided into ever smaller particles or whether eventually an indivisible particle would be encountered. Democritus in about 450 B.C. argued that substances were ultimately composed of small, indivisible particles that he labeled atoms. He further suggested that different substances were composed of different atoms or combinations of atoms, and that one substance could be converted into another by rearranging the atoms. It was impossible to conclusively prove or disprove this proposal for more than 2000 years.

The modern proof for the atomic nature of matter was first proposed by the English chemist John Dalton in 1803. Dalton stated that each chemical element possesses a particular kind of atom, and any quantity of the element is made up of identical atoms of this kind. What distinguishes one element from another element is the kind of atom of which it consists, and the basic physical difference between kinds of atoms is their weight.

## SUBATOMIC PARTICLES

For almost 100 years after Dalton established the atomic nature of atoms, it was considered impossible to divide the atom into even smaller parts. All of the results of chemical experiments during this time indicated that the atom was indivisible. Eventually, experimentation into electricity and radioactivity indicated that particles of matter smaller than the atom did indeed exist. In 1906, J. J. Thompson won the Nobel Prize in physics for establishing the existence of electrons. Electrons are negatively-charged particles that have 1/1835 the mass of the hydrogen atom. Soon after the discovery of electrons, protons were discovered. Protons are relatively large particles that have almost the same mass as a hydrogen atom and a positive charge equal in magnitude (but opposite in sign) to that of the electron. The third subatomic particle to be discovered, the neutron, was not found until 1932. The neutron has almost the same mass as the proton, but it is electrically neutral.

### The Atom

By the early 20th century, there was rather compelling evidence that matter could be described by an atomic theory. That is, matter is composed of relatively few building blocks that we refer to as atoms. This theory provided a consistent and unified picture for all known chemical processes at that time. However, some mysteries could not be explained by this atomic theory. In 1896, A.H. Becquerel discovered penetrating radiation. In 1897, J.J. Thomson showed that electrons have negative electric charge and come from ordinary matter. For matter to be electrically neutral, there must also be positive charges lurking somewhere. Where are and what carries these positive charges?

A monumental breakthrough came in 1911 when Ernest Rutherford and his coworkers conducted an experiment intended to determine the angles through which a beam of alpha particles (helium nuclei) would scatter after passing through a thin foil of gold.

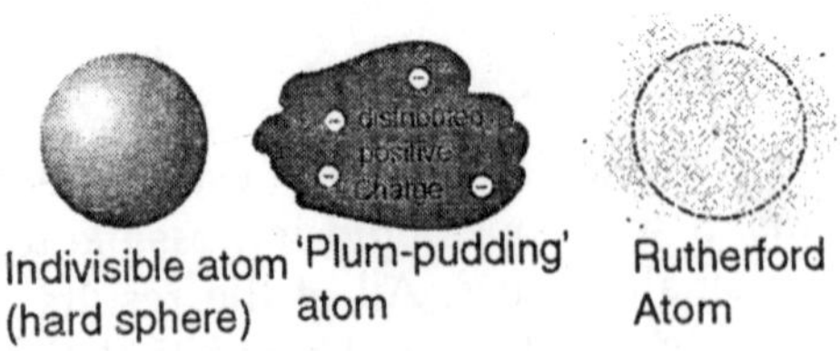


Models of the atom. The dot at the centre of the Rutherford atom is the nucleus.

What results would be expected for such an experiment? It depends on how the atom is organized. A prevailing model of the atom at the time (the Thomson, or "plum-pudding," atom) proposed that the negatively charged electrons (the plums) were mixed with smeared-out positive charges (the pudding). This model explained the neutrality of bulk material, yet still allowed the description of the flow of electric charges. In this model, it would be very unlikely for an alpha particle to scatter through an angle greater than a small fraction of a degree, and the vast majority should undergo almost no scattering at all.

The results from Rutherford's experiment were astounding. The vast majority of alpha particles behaved as expected, and hardly scattered at all. But there were alpha particles that scattered through angles greater than 90 degrees, incredible in light of expectations for a "plum-pudding" atom. It was largely the evidence from this type of experiment that led to the model of the atom as having a nucleus. The only model of the atom consistent with this Rutherford experiment is that a small central core (the nucleus) houses the positive charge and most of the mass of the atom, while the majority of the atom's volume contains discrete electrons orbiting about the central nucleus.

Under classical electromagnetic theory, a charge that is moving in a circular path, loses energy. In Rutherford's model, the electrons orbit the nucleus similar to the orbit of planets about the sun. However, under this model, there is nothing to prevent the electrons from losing energy and falling into the nucleus under the influence of its Coulomb attraction. This stability problem was solved by Niels Bohr in 1913 with a new model in which there are particular orbits in which the electrons do not lose energy and therefore do not spiral into the nucleus. This model was the beginning of *quantum mechanics,* which successfully explains many properties of atoms. Bohr's model of the atom is still a convenient description of the energy levels of the hydrogen atom.

## Bohr Model of the Atom

The British physicist Ernest Rutherford postulated that the positive charge in an atom is concentrated in a small region called a nucleus at the centre of the atom with electrons existing in orbits around it. Niels Bohr, coupling Rutherford's postulation with the quantum theory introduced by Max Planck, proposed that the atom consists of a dense nucleus of protons surrounded by electrons traveling in discrete orbits at fixed distances from the nucleus. An electron in one of these orbits or shells has a specific or discrete quantity of energy (quantum). When an electron moves from one allowed orbit to another allowed orbit, the energy difference between the two states is emitted or absorbed in the form of a single quantum of radiant energy called a photon. Bohr's model of the hydrogen atom showing an electron as having just dropped from the third shell to the first shell with the emission of a photon that has an energy = $\hbar$ n. ($\hbar$ = Planck's constant = 6.63 x $10^{-34}$ J-s and n = frequency of the photon.) Bohr's theory was the first to successfully account for the discrete energy levels of this radiation as measured in the laboratory. Although Bohr's atomic model is designed specifically to explain the hydrogen atom, his theories apply generally to the structure of all atoms.

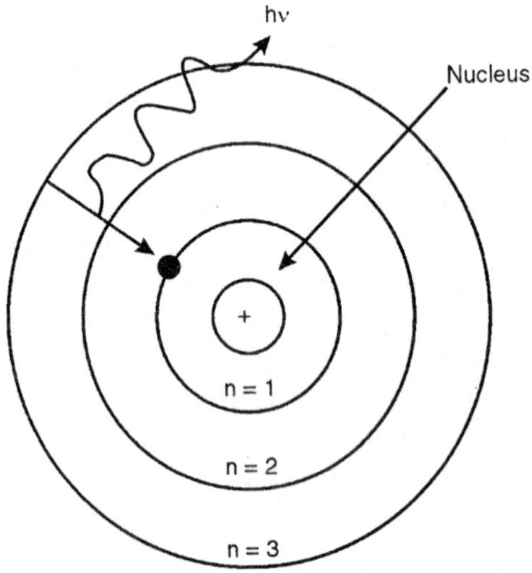


**Properties of Subatomic Particles**

| Particle | Location | Charge | Mass |
|---|---|---|---|
| Neutron | Nucleus | none | 1.008665 amu |
| Proton | Nucleus | +1 | 1.007277 amu |
| Electron | Shells around nucleus | -1 | 0.0005486 am |

## MEASURING UNITS ON THE ATOMIC SCALE

The size and mass of atoms are so small that the use of normal measuring units, while possible, is often inconvenient. Units of measure have been defined for mass and energy on the atomic scale to make measurements more convenient to express. The unit of measure for mass is the atomic mass unit (amu). One atomic mass unit is equal to $1.66 \times 10^{-24}$ grams. The reason for this particular value for the atomic mass unit will be discussed in a later chapter. The mass of a neutron and a proton are both about 1 amu. The unit for energy is the electron volt (eV). The electron volt is the amount of energy acquired by a single electron when it falls through a potential difference of one volt. One electron volt is equivalent to

$$1.602 \times 10^{-19} \text{ joules or } 1.18 \times 10^{-19} \text{ foot-pounds.}$$

## The Nucleus

The nucleus depicted above is now understood to be a quantum system composed of protons and neutrons, particles of nearly equal mass and the same intrinsic angular momentum (spin) of 1/2. The proton carries one unit of positive electric charge while the neutron has no electric charge. The term nucleon is used for either a proton or a neutron. The simplest nucleus is that of hydrogen, which is just a single proton, while the largest nucleus studied has nearly 300 nucleons. A nucleus is identified as in the example below by its atomic number $Z$ (*i.e.*, the number of protons), the neutron number, $N$, and the mass number, $A$, where $A = Z + N$.

A mass number

Z atomic number $\quad {}^{14}_{6}C$

N neutron number = A –Z

The convention for designating nuclei is by atomic number, $Z$, and mass number, $A$, as well as its chemical symbol. The neutron number is given by $N = A–Z$.

In addition to its atomic number and mass number, a nucleus is also characterized by its size, shape, binding energy, angular momentum, and (if it is unstable) half-life. One of the best ways to determine the size of a nucleus is to scatter high-energy electrons from it. The angular distribution of the scattered electrons depends on the proton distribution. The proton distribution can be characterized by an average radius. It is found that nuclear radii range from 1 – 10′ $10^{-15}$ m. This radius is much smaller than that of the atom, which is typically $10^{-10}$ m. Thus, the nucleus occupies an extremely small volume inside the atom. The nuclei of some atoms are spherical, while others are stretched or flattened into deformed shapes.

The binding energy of a nucleus is the energy holding a nucleus together. This energy varies from nucleus to nucleus and increases as *A* increases. Because of variations in binding energy, some nuclei are unstable and decay into other ones. The rate of decay is related to the mean lifetime of the decaying nucleus. The time required for half of a population of unstable nuclei to decay is called the half-life. Half-lives vary from tiny fractions of a second to billions of years.

## Atomic and Nuclear Radii

The size of an atom is difficult to define exactly due to the fact that the electron cloud, formed by the electrons moving in their various orbitals, does not have a distinct outer edge. A reasonable measure of atomic size is given by the average distance of the outermost electron from the nucleus. Except for a few of the lightest atoms, the average atomic radii are approximately the same for all atoms, about $2\times10^{-8}$ cm. Like the atom the nucleus does not have a sharp outer boundary. Experiments have shown that the nucleus is shaped like a sphere with a radius that depends on the atomic mass number of the atom. The relationship between the atomic mass number and the radius of the nucleus is shown in the following equation.

$$r = (1.25 \times 10^{-13} \text{ cm})\, A^{1/3}$$

where:

r = radius of the nucleus (cm)

A = Atomic mass number (dimensionless)

## NUCLEAR FORCES

In the Bohr model of the atom, the nucleus consists of positively-charged protons and electricallyneutral neutrons. Since both protons and neutrons exist in the nucleus, they are both referred to as nucleons. One problem that the Bohr model of the atom presented was accounting for an attractive force to overcome the repulsive force between protons. Two forces present in the nucleus are (1) electrostatic forces between charged particles and (2) gravitational forces between any two objects that have mass. It is possible to calculate the magnitude of the gravitational force and electrostatic force based upon principles from classical physics.

Newton stated that the *gravitational force* between two bodies is directly proportional to the masses of the two bodies and inversely proportional to the square of the distance between the bodies.

$$F_g = \frac{Gm_1m_2}{r^2}$$

Where

$F_g$ = gravitational force (newtons)

$m_1$ = mass of first body (kilograms)

$m_2$ = mass of second body (kilograms)

G = gravitational constant ($6.67 \times 10^{-11}$ N–$m^2/kg^2$

The equation illustrates that the larger the masses of the objects or the smaller the distance between the objects, the greater the gravitational force. So even though the masses of nucleons are very small, the fact that the distance between nucleons is extremely short may make the gravitational force significant. It is necessary to calculate the value for the gravitational force and compare it to the value for other forces to determine the significance of the gravitational force in the nucleus. The gravitational force between two protons that are separated by a distance of 10–20 meters is about 10–24 newtons.

Coulomb's Law can be used to calculate the force between two protons. The *electrostatic force* is directly proportional to the electrical charges of the two particles and inversely proportional to the square of the distance between the particles. Coulomb's Law is stated as the following equation.

$$F_e = \frac{KQ_1Q_2}{r^2}$$

where

$F_e$ = electrostatic force (newtons)

K = electrostatic constant ($9.0 \times 10^9$ N–m$^2$/C$^2$)

$Q_1$ = charge of first particle (colombs)

$Q_2$ = charge of second particle (colombs)

r = distance betwen particles (meters)

Using this equation, the electrostatic force between two protons that are separated by a distance of 10–20 meters is about $10^{12}$ newtons. Comparing this result with the calculation of the gravitational force ($10^{-24}$ newtons) shows that the gravitational force is so small that it can be neglected. If only the electrostatic and gravitational forces existed in the nucleus, then it would be impossible to have stable nuclei composed of protons and neutrons. The gravitational forces are much too small to hold the nucleons together compared to the electrostatic forces repelling the protons. Since stable atoms of neutrons and protons do exist, there must be another attractive force acting within the nucleus. This force is called the nuclear force. The *nuclear force* is a strong attractive force that is independent of charge. It acts equally only between pairs of neutrons, pairs of protons, or a neutron and a proton. The nuclear force has a very short range; it acts only over distances approximately equal to the diameter of the nucleus ($10^{-13}$ cm). The attractive nuclear force between all nucleons drops off with distance much faster than the repulsive electrostatic force between protons.

## Isotopes

*Isotopes* are nuclides that have the same atomic number and are therefore the same element, but differ in the number of neutrons. Most elements have a few stable isotopes and several unstable, radioactive isotopes. Oxygen has three stable isotopes that can be found in nature (oxygen–16, oxygen–17, and oxygen–18) and eight radioactive isotopes. Another example is hydrogen, which has two stable isotopes (hydrogen–1 and hydrogen–2) and a single radioactive isotope (hydrogen-3). The isotopes of hydrogen are unique in that they are each commonly referred to by a unique name instead of the common chemical element name. Hydrogen-1 is almost always referred to as hydrogen, but the term protium is infrequently used also. Hydrogen–2 is commonly called deuterium and symbolized $^{2}_{1}D$. Hydrogen-3 is commonly called tritium and symbolized $^{3}_{1}T$. This text will normally use the symbology $^{2}_{1}H$ and $^{2}_{1}H$ for deuterium and tritium, respectively.

## The Isotopes of Hydrogen

It is often useful to study the simplest system. Therefore, hydrogen, the simplest nucleus, has been studied extensively. The isotopes of hydrogen show many of the effects found in more complicated nuclei. (The word isotope refers to a nucleus with the same $Z$ but different $A$).

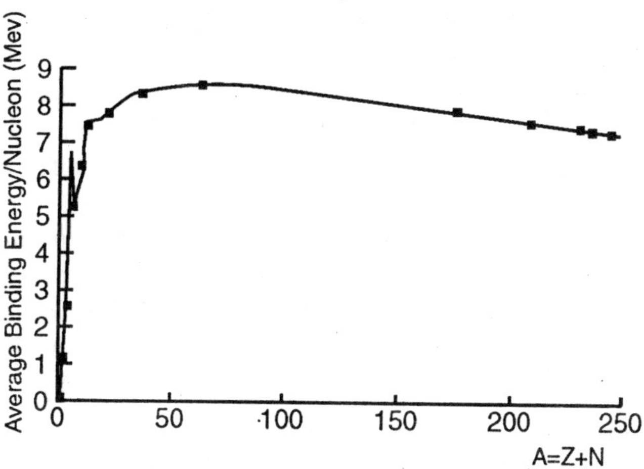


**Fig.** The Curve of The Average Binding Energy Per Nucleon.

There are three isotopes of the element hydrogen: hydrogen, deuterium, and tritium. How do we distinguish between them? They each have one single proton ($Z$=1), but differ in the number of their neutrons. Hydrogen has no neutron, deuterium has one, and tritium has two neutrons. The isotopes of hydrogen have, respectively, mass numbers of one, two, and three. Their nuclear symbols are therefore $^{1}H$, $^{2}H$, and $^{3}H$. The atoms of these isotopes have one electron to balance the charge of the one proton. Since chemistry depends on the interactions of protons with electrons, the chemical properties of the isotopes are nearly the same.

Energy may be released as a packet of electromagnetic radiation, a photon. Photons created in nuclear processes are labeled gamma rays (denoted by the Greek letter gamma, $\gamma$. When a proton and neutron combine to form deuterium, the reaction can be written $^{1}n + ^{1}H \rightarrow ^{2}H + \gamma$. Energy must balance in this equation. Mass can be written in atomic mass units (u) or in the equivalent energy units of million electron-volts divided by the square of the speed of light (MeV)/$c^2$. (From Einstein's mass-energy equivalence equation, $E = mc^2$, u = 931.5 MeV/$c^2$.) The mass of the deuterium nucleus (2.01355 u) is less than the sum of the masses of the proton (1.00728 u) and the neutron (1.00866 u), which is 2.01594 u. Where has the missing mass (0.00239 u) gone? The answer is that the attractive nuclear force between the nucleons has created a negative nuclear potential energy–the binding energy $E_B$– that is related to the missing mass, $\Delta m$ (the difference between the two masses). The photon released in forming deuterium has an energy of 2.225 MeV, equivalent to the 0.00239 u required to separate the proton and neutron back into unbound particles. The nuclear decay photons are, in general, higher in energy than photons created in atomic processes.

When tritium is formed by adding a neutron to deuterium, $^{1}n + ^{2}H \rightarrow ^{3}H + \gamma$, a larger amount of energy is released–6.2504 MeV. The greater binding energy of tritium compared to deuterium shows that the nuclear potential energy does not grow in a simple way with the addition of nucleons (the total binding energy is roughly proportional to A). The binding energy per nucleon continues to grow as protons and neutrons are added to construct more massive nuclei until a maximum of about 8 MeV per nucleon is reached around A = 60, past which the average binding energy per nucleon slowly decreases up to the most massive nuclei, for which it is about 7 MeV.

How does a nucleus, which can have up to approximately 100 protons, hold itself together? Why does the electrical repulsion among all those positive charges not cause the nucleus to break up? There must be an attractive force strong enough to be capable of overcoming the repulsive Coulomb forces between protons. Experiment and theory have come to recognize an attractive nuclear interaction that acts between nucleons when they are close enough together (when the range is short enough). The balance between electromagnetic and nuclear forces sets the limit on how large a nucleus can grow.

## Natural Abundance of Isotopes

The relative abundance of an isotope in nature compared to other isotopes of the same element is relatively constant. The Chart of the Nuclides presents the relative abundance of the naturally occurring isotopes of an element in units of atom percent. Atom percent is the percentage of the atoms of an element that are of a particular isotope. Atom percent is abbreviated as a/o. If a cup of water contains $8.23 \times 10^{24}$ atoms of oxygen, and the isotopic abundance of oxygen-18 is 0.20percent, then there are $1.65 \times 10^{22}$ atoms of oxygen-18 in the cup.

The atomic weight for an element is defined as the average atomic weight of the isotopes of the element. The atomic weight for an element can be calculated by summing the products of the isotopic abundance of the isotope with the atomic mass of the isotope.

## Enriched and Depleted Uranium

Natural uranium mined from the earth contains the isotopes uranium–238, uranium–235 and uranium–234. The majority (99.2745percent) of all the atoms in natural uranium are uranium–238. Most of the remaining atoms (0.72percent) are uranium–235, and a slight trace (0.0055percent) are uranium–234. Although all isotopes of uranium have similar chemical properties, each of the isotopes has significantly different nuclear properties. For reasons that will be discussed in later modules, the isotope uranium–235 is usually the desired material for use in reactors. A vast amount of equipment and energy are expended in processes that separate the isotopes of uranium (and other elements).

The details of these processes are beyond the scope of this module. These processes are called enrichment processes because they selectively increase the proportion of a particular isotope. The enrichment process typically starts with feed material that has the proportion of isotopes that occur naturally. The process results in two types of In the case of uranium, the natural uranium ore is 0.72 a/o uranium–235. The desired outcome of the enrichment process is to produce enriched uranium. *Enriched uranium* is defined as uranium in which the isotope uranium–235 has a concentration greater than its natural value. The enrichment process will also result in the byproduct of depleted uranium. *Depleted uranium* is defined as uranium in which the isotope uranium–235 has a concentration less than its natural value. Although depleted uranium is referred to as a by-product of the enrichment process, it does have uses in the nuclear field and in commercial and defence industries.

## Theoretical Models

A goal of nuclear physics is to account for the properties of nuclei in terms of mathematical models of their structure and internal motion. Three important nuclear models are the Liquid Drop Model, the Shell Model, which emphasizes the orbits of individual nucleons in the nucleus, and the *Collective Model,* which complements the shell model by including motions of the whole nucleus such as rotations and vibrations.

The Liquid Drop Model treats the nucleus as a liquid. Nuclear properties, such as the binding energy, are described in terms of volume energy, surface energy, compressibility, etc.–parameters that are usually associated with a liquid. This model has been successful in describing how a nucleus can deform and undergo fission.

The Nuclear Shell Model is similar to the atomic model where electrons arrange themselves into shells around the nucleus. The least-tightly-bound electrons (in the incomplete shells) are known as valence electrons because they can participate in exchange or rearrangement, that is, chemical reactions. The shell structure is due to the quantum nature of electrons and the fact that electrons are *fermions*–particles of half-integer spin. Particles with integer spin are *bosons*. A group of bosons all tend to occupy the same state (usually the state with the lowest energy), whereas fermions with the same quantum numbers do just the opposite: they avoid each other. Consequently the fermions in a bound system will gradually fill up the available states: the lowest one first, then the next higher unoccupied state, and so on up to the valence shell. In atoms, the electrons obey the *Pauli Exclusion Principle*, which is responsible for the observed number of electrons in each possible state (at most 2) characterized by quantum numbers $n$, $l$, and $m$. It is the Pauli Principle (based on the fermionic nature of electrons) that gives the periodic structure to both atomic and nuclear properties.

Since protons and neutrons are also fermions, the energy states the nucleons occupy are filled from the lowest to the highest as nucleons are added to the nucleus. In the shell model the nucleons fill each energy state with nucleons in orbitals with definite angular momentum. There are separate energy levels for protons and neutrons. The ground state of a nucleus has each of its protons and neutrons in the lowest possible energy level. Excited states of the nucleus are then described as promotions of nucleons to higher energy levels. This model has been very successful in explaining the basic nuclear properties. As is the case with atoms, many nuclear properties (angular momentum, magnetic moment, shape, etc.) are dominated by the last filled or unfilled valence level.

The Collective Model emphasizes the coherent behaviour of all of the nucleons. Among the kinds of collective motion that can occur in nuclei are rotations or vibrations that involve the entire nucleus. In this respect, the nuclear properties can be analysed using the same description that is used to analyse the properties of a charged drop of liquid suspended in space. The Collective Model can thus be viewed as an extension of the Liquid Drop Model; like the Liquid Drop Model, the Collective Model provides a good starting point for understanding fission.

In addition to fission, the Collective Model has been very successful in describing a variety of nuclear properties, especially energy levels in nuclei with an even number of protons and neutrons. These even nuclei can often be treated as having no valence particles so that the Shell Model does not apply. These energy levels show the characteristics of rotating or vibrating systems expected from the laws of quantum mechanics. Commonly measured properties of these nuclei, including broad systematics of excited state energies, angular momentum, magnetic moments, and nuclear shapes, can be understood using the Collective Model.

The Shell Model and the Collective Model represent the two extremes of the behaviour of nucleons in the nucleus. More realistic models, known as unified models, attempt to include both shell and collective behaviours.

## Sub-nucleonic Structure and the Modern Picture of a Nucleus

Do protons and neutrons have internal structure? The answer is yes. With the development of higher and higher energy particle accelerators, physicists have found experimentally that the nucleons are complex objects with their own interesting internal structures.

One of the most significant developments in modern physics is the emergence of the Standard Model of Fundamental Interactions. This model states that the material world is made up of two categories of particles, quarks and leptons, together with their antiparticle counterparts. The leptons are either neutral (such as the neutrino) or carry one unit of charge, *e* (such as the electron, muon, and tau). The quarks are pointlike objects with charge 1/3*e* or 2/3 *e*. Quarks are spin–1/2 particles, and therefore are fermions, just as electrons are.

The quarks and leptons can be arranged into three families. The up- and down-quarks with the electron and the electron neutrino form the family that makes up ordinary matter. The other two families produce particles that are very short-lived and do not significantly affect the nucleus. It is a significant fact in the evolution of the universe that only three such families are found in nature–more families would have lead to a quite different world.

## The Standard Model of Particles and Interactions

One could imagine, then, trying to understand the structure of protons and neutrons in terms of the fundamental particles described in the Standard Model. Because the protons and neutrons of ordinary matter are affected by the strong interaction (i.e., the interaction that binds quarks and that ultimately holds nuclei together), they fall into the category of composite particles known as hadrons. Hadrons that fall into the subcategory known as baryons are made of three quarks. Protons, which consist of two up and one down quark, and neutrons (two down and one up quark) are baryons. There are also hadrons called mesons, which are made of quark-antiquark pairs, an example of which is the pion.

Because baryons and mesons have internal quark structure, they can be put into excited states, just as atoms and nuclei can. This requires that energy be deposited in them. One example is the first excited state of the proton, usually referred to as the Delta–1232 (where 1232 MeV/$c^2$ is the mass of the particle). In the Delta, it is thought that one of the quarks gains energy by flipping its spin with respect to the other two. In an atom, the energy needed to excite an electron to a higher state is on the order of a few to a thousand electron volts. In comparison, in a nucleus, a single nucleon excitation typically costs a MeV ($10^6$ eV). In a proton, it takes about 300 MeV to flip the spin of a quark. This kind of additional energy is generally only available by bombarding the proton with energetic particles from an accelerator.

## MASS DEFECT AND BINDING ENERGY

### Mass Defect

Careful measurements have shown that the mass of a particular atom is always slightly less than the sum of the masses of the individual neutrons, protons, and electrons of which the atom consists. The difference between the mass of the atom and the sum of the masses of its parts is called the mass defect ($\Delta m$). In calculating the mass defect it is important to use the full accuracy of mass measurements because the difference in mass is small compared to the mass of the atom. Rounding off the masses of atoms and particles to three or four significant digits prior to the calculation will result in a calculated mass defect of zero.

$$\Delta m = [Z(m_p + m_e) + (A-Z)m_n] - m_{atom}$$

where:

$\Delta m$ = mass defect (amu)

mp = mass of a proton (1.007277 amu)

mn = mass of a neutron (1.008665 amu)

me = mass of an electron (0.000548597 amu)

matom = mass of nuclide $^{A}Z$ X (amu)

Z = atomic number (number of protons)

A = mass number (number of nucleons)

### Binding Energy

The loss in mass, or mass defect, is due to the conversion of mass to binding energy when the nucleus is formed. *Binding energy* is defined as the amount of energy that must be supplied to a nucleus to completely separate its nuclear particles (nucleons). It can also be understood as the amount of energy that would be released if the nucleus was formed from the separate particles. Binding energy is the energy equivalent of the mass defect. Since the mass defect was converted to binding energy (BE) when the nucleus was formed, it is possible to calculate the binding energy using a conversion factor derived by the mass-energy relationship from Einstein's Theory of Relativity. Einstein's famous equation relating mass and energy is $E = mc^2$ where c is the velocity of light ($c = 2.998 \times 108$ m/sec). The energy equivalent of 1 amu can be determined by inserting this quantity of mass into Einstein's equation and applying conversion factors.

$$E = mc^2$$

$$= 1\text{amu}\left(\frac{1.6606\times10^{-27}}{1\text{amu}}\right)\left(2.998\times10^{8}\frac{m}{sec}\right)^2$$

$$\left(\frac{1\text{N}}{1\frac{\text{kg}^{-m}}{\text{sec}^2}}\right)\left(\frac{1\text{J}}{1\text{N}^{-m}}\right)$$

$$= 1.4924\times10^{-10}J\left(\frac{1Mev}{1.6022\times10^{-13}J}\right)$$

$$s = 93.5\ Mev$$

Conversion factors:

1 asmu = $1.6606\times10^{-27}kg$

1 newton = $1\ kg\text{-}m/sec^2$

1 *joule* = 1 *newton-meter*

1 MeV = $1.6022 \times 10^{-13}$ *joules*

Since 1 amu is equivalent to 931.5 MeV of energy, the binding energy can be calculated using Equation

$$B.E. = \Delta m\left(\frac{931.5\text{MeV}}{1\text{ amu}}\right)$$

Example:

Calculate the mass defect and binding energy for uranium – 235. Oneuranum –235 atom has amass of 235.043924 amu.

Solution:

Step 1: Calculate the mass defect using Equation.

$$\Delta m = \left[Z\left(m_p + m_e\right) + (1 - Z)m_n\right] - m_{atom}$$

$$\Delta m = [92\left(1.007826\,\text{amu}\right) + (235 - 92)$$

$$1.008665\,\text{amu}] - 235.043924\,\text{amu}$$

$$\Delta m = 1.91517 \text{ amu}$$

step 2: Use the mass defect and Equation to calcualte the binding energy.

$$\text{B.E.} = \Delta\text{m}\left(\frac{931.5\text{MeV}}{1\text{amu}}\right)$$

$$= 1.91517\,\text{amu}\left(\frac{931.5\text{MeV}}{1\text{amu}}\right)$$

$$= 1764 \text{ MeV}$$

**Energy Levels of Atoms**

The electrons that circle the nucleus move in fairly well-defined orbits. Some of these electrons are more tightly bound in the atom than others. Only 7.38 eV is required to remove the outermost electron from a lead atom, while 88,000 eV is required to remove the innermost electron. The process of removing an electron from an atom is called ionization, and the energy required to remove the electron is called the ionization energy. In a neutral atom (number of electrons = Z) it is possible for the electrons to be in a variety of different orbits, each with a different energy level.

The state of lowest energy is the one in which the atom is normally found and is called the ground state. When the atom possesses more energy than its ground state energy, it is said to be in an excited state. An atom cannot stay in the excited state for an indefinite period of time. An excited atom will eventually transition to either a lower-energy excited state, or directly to its ground state, by emitting a discrete bundle of electromagnetic energy called an x-ray. The energy of the x-ray will be equal to the difference between the energy levels of the atom and will typically range from several eV to 100,000 eV in magnitude.

**Energy Levels of the Nucleus**

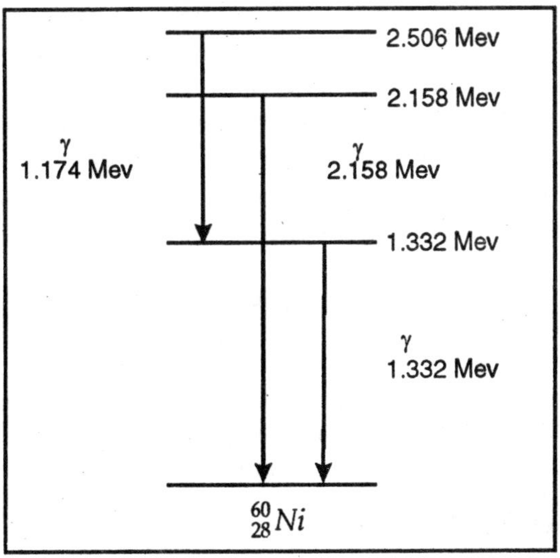


The nucleons in the nucleus of an atom, like the electrons that circle the nucleus, exist in shells that correspond to energy states. The energy shells of the nucleus are less defined and less understood than those of the electrons. There is a state of lowest energy (the ground state) and discrete possible excited states for a nucleus. Where the discrete energy states for the electrons of an atom are measured in eV or keV, the energy levels of the nucleus are considerably greater and typically measured in MeV. A nucleus that is in the excited state will not remain at that energy level for an indefinite period. Like the electrons in an excited atom, the nucleons in an excited nucleus will transition towards their lowest energy configuration and in doing so emit a discrete bundle of electromagnetic radiation called a gamma ray ($\gamma$-ray). The only differences between x-rays and g-rays are their energy levels and whether they are emitted from the electron shell or from the nucleus.

The ground state and the excited states of Energy Level Diagram - Nickel-60 a nucleus can be depicted in a nuclear energy-level diagram. The nuclear energy-level diagram consists of a stack of horizontal bars, one bar for each of the excited states of the nucleus. The vertical distance between the bar representing an excited state and the bar representing the ground state is proportional to the energy level of the excited state with respect to the ground state. This difference in energy between the ground state and the excited state is called the excitation energy of the excited state. The ground state of a nuclide has zero excitation energy. The bars for the excited states are labeled with their respective energy levels.

## QUANTUM NUMBERS

As in any quantum system there are many quantum states in each nucleus. These are labelled by their quantum numbers, which, as will be shown later, originate in symmetries of the underlying Hamiltonian, or rather the underlying physics.

### ANGULAR MOMENTUM

One of the key invariances of the laws of physics is rotational invariance, i.e., physics is independent of the direction you are looking at. This leads to the introduction of a vector angular momentum operator,

$$\hat{L} = \hat{r} \times \hat{p},$$

which generates rotations. Quantum states are not necessarily invariant under the rotation, but transform in a well-defined way. The three operators $\hat{L}x, \hat{L}y$, and $\hat{L}x$ satisfy a rather intriguing structure,

$$[\hat{L}x, \hat{L}y] = \hat{L}x\hat{L}y - \hat{L}y\hat{L}x = i\hbar\hat{L}x,$$

and the same for q cyclic permutation of indices ($xyz \rightarrow yzx$ or $zxy$). This shows that we cannot determine all three components simultaneously in a quantum state. One normally only calculates the length of the angular momentum vector, and its projection on the z axis,

$$\hat{L}^2 \phi LM = \hbar^2 L(L+1)$$

$$\hat{L}_z \phi LM = \hbar Lz \phi LM$$

It can be shown that $L$ is a non-negative integer, and $M$ is an integer satisfying $|M| < L$, i.e., the projection is always smaller than or equal to the length, a rather simple statement in classical mechanics.

The standard, albeit slightly simplified, picture of this process is that of a fixed length angular momentum precessing about the z axis, keeping the projection fixed.

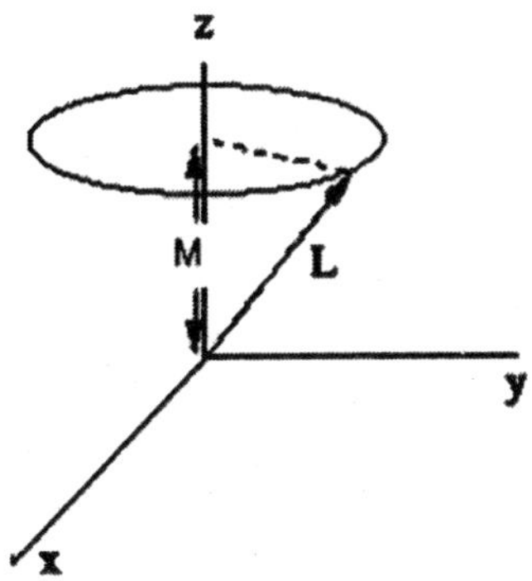


**Fig.** A pictorial representation of the "quantum precession" of an angular momentum of fixed length $L$ and projection $M$.

The energy of a quantum state is independent of the $M$ quantum number, since the physics is independent of the orientation of $L$ in space (unless we apply a magnetic field that breaks this symmetry). We just find multiplets of $2L + 1$ states with the same energy and value of $L$, differing only in $M$.

Unfortunately the story does not end here. Like electrons, protons and neutron have a spin, i.e., we can use a magnetic field to separate nucleons with spin up from those with spin down. Spins are like orbital angular momenta in many aspects, we can write three operators $\hat{S}$ that satisfy the same relation as the $\hat{L}$'s, but we find that

$$\hat{S}^2 \phi S, SSz = \hbar^2 \frac{3}{4} \phi S, Sz$$

i.e., the length of the spin is 1/2, with projections ±1/2.

Spins will be shown to be coupled to orbital angular momentum to total angular momentum *J*,

$$\hat{J} = \hat{L} + \hat{S},$$

and we specify the quantum state by *L*, *S*, *J* and $J_z$.

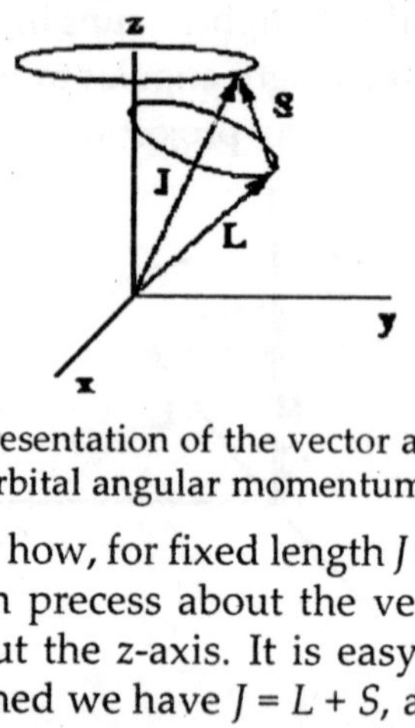


**Fig.** A pictorial representation of the vector addition of spin and orbital angular momentum.

There we show how, for fixed length *J* the spin and orbital angular momentum precess about the vector J, which in its turn precesses about the z-axis. It is easy to see that if *vecL* and S are fully aligned we have $J = L + S$, and if they are anti-aligned $J = | L - S |$. A deeper quantum analysis shows that this is the way the quantum number work. If the angular momentum quantum numbers of the states being coupled are *L* and *S*, the length of the resultant vector *J* can be

$$J = | L - S |, | L - S | + 2, \ldots, L + S.$$

We have now discussed the angular momentum quantum number for a single particle. For a nucleus which in principle is made up from many particles, we have to add all these angular momenta together until we get something called the total angular momentum. Since the total angular momentum of a single particle is half-integral, the total angular momentum of a nucleus is integer for even $A$, and half-integer for odd $A$.

## PARITY

Another symmetry of the wave function is parity. If we change $r \rightarrow r$, i.e., mirror space, the laws of physics are invariant. Since we can do this operation twice and get back where we started from, any eigenvalue of this operation must be ±1, usually denoted as $\Pi = \pm$. It can be shown that for a particle with orbital angular momentum $L$, $\Pi = (-1)^L$. The parity of many particles is just the product of the individual parities.

## ISOTOPIC SPIN (ISOBARIC SPIN, ISOSPIN)

The most complicated symmetry in nuclear physics is isospin. In contrast to the symmetries above this is not exact, but only approximate. The first clue of this symmetry come from the proton and neutron masses, $m_n$ = MeV/$c^2$ and $m_p$ = MeV/$c^2$, and their very similar behaviour in nuclei. Remember that the dominant binding terms only depended on the number of nucleons, not on what type of nucleons we are dealing with.

All of this leads to the assumption of another abstract quantity, called isospin, which describes a new symmetry of nature. We assume that both neutrons and protons are manifestation of one single particle, the nucleon, with isospin down or up, respectively. We have to see whether this makes sense by looking in more detail at the nuclear physics. We propose the identification

$$Q = (I_z + 1/2)e,$$

where $I_z$ is the z projection of the vectorial quantity called isospin. Apart from the neutron-proton mass difference, isospin symmetry in nuclei is definitely broken by the Coulomb force, which acts on protons but not on neutrons. We argue that the nuclear force, that couples to the "nucleon charge" rather than electric charge, respects this symmetry. Look at a few nuclei where we can study both a nucleus and its mirror image under the exchange of protons and neutrons. One example are the nuclei $^7He$ and $^7B$ (2 protons and 5 neutrons, $I_z$ = - 3/2 vs. 5 protons and 2 neutrons, $I_z$ = 3/2) and $^7Li$ and $^7B$ (3 protons and 4 neutrons, $I_z$ = - 1/2 vs. 4 protons and 3 neutrons, $I_z$ = 1/2).

We note there the great similarity between the pairs of mirror nuclei. Of even more importance is the fact that the $3/2^-;3/2$ level occurs at the same energy in all four nuclei, suggestion that we can define these states as an "isospin multiplet", the same state just differing by $I_z$

## DEUTERON

Let us think of the deuteron (initially) as a state with $L$ = 0, $J$ = 1, $S$ = 1, usually denoted as $^3S_1$ ($S$ means $L$ = 0, the 3 denotes $S$ = 1, i.e., three possible spin orientations, and the subscript 1 the value of $J$). Let us model the nuclear force as a three dimensional square well with radius $R$. The Schrodinger equation for the spherically symmetric $S$ state is (work in radial coordinates)

$$-\frac{\hbar^2}{2\mu}\frac{1}{r^2}\left(\frac{d}{dr}r^2\frac{d}{dr}\right) R(r) + V(r)R(r) = ER(r).$$

Here $V(r)$ is the potential, and μ is the reduced mass,

$$\mu = \frac{m_n m_p}{m_n + m_p},$$

which arrises from working in the relative coordinate only. It is easier to work with $u(r) = rR(r)$, which satisfies the condition

$$\frac{\hbar}{2\mu}\frac{d^2}{dr^2} - u(r) + V(r)u(r) = Eu(r),$$

as well as $u(0) = 0$. The equation in the interior

$$-\frac{\hbar}{2\mu}\frac{d^2}{dr^2}\ u(r)V_0u(r) = Eu(r), \qquad u(0) = 0$$

has as solution

$$u = A\ \sin kr, \qquad k = \sqrt{\frac{2\mu}{\hbar^2}(V_0 + E)}\ .$$

Outside the well we find the standard damped exponential,

$$u = B\ \exp(kr), \qquad k = \sqrt{\frac{2\mu}{\hbar^2}(-E)}\ .$$

Matching derivatives at the boundary we find

$$-\cot kR = \frac{K}{k} = \sqrt{\frac{-E}{V_0 + E}}\ .$$

Now make the assumption that $|E| \ll V_0$, which will prove true. Then we find

$$k \approx \sqrt{\frac{2\mu}{\hbar}V_0}\ , \qquad \cot k\,R \approx 0.$$

Since it is known from experiment that the deuteron has only one bound state at energy -2.224573±0.000002 MeV, we see that $kR \approx \pi/2$! Substituting $k$ we see that

$$V_0R^2 = \frac{\pi^2\hbar^2}{8\mu}\ .$$

If we take $V_0 = 30$ MeV, we find $R = 1.83$ fm.

We can orient the spins of neutron and protons in a magnetic field, i.e., we find that there is an energy

$$E_{magn} = \mu\ N\ \mu\ S \cdot B.$$

(The units for this expression is the so-called nuclear magneton, $\mu N = \frac{e\hbar}{2m_p}$.) Experimentally we know that

$$\mu_n = -\ 1.91315 \pm 0.00007\mu_N \qquad \mu_p = 2.79271 \pm 0.00002\ \ \mu_N$$

If we compare the measured value for the deuteron, $\mu_d$= 0.857411±0.000019$\mu_N$, with the sum of protons and neutrons (spins aligned), we see that $\mu_p$+ $\mu_n$= 0.857956±0.00007$\mu_N$. The close agreement suggest that the spin assignment is largely OK; the small difference means that our answer cannot be the whole story: we need other components in the wave function.

We know that an *S* state is spherically symmetric and cannot have a quadrupole moment, i.e., it does not have a preferred axis of orientation in an electric field. It is known that the deuteron has a positive quadrupole moment of 0.29$e^2$ fm$^2$, corresponding to an elongation of the charge distribution along the spin axis.

From this we conclude that the deuteron wave function carries a small (7 per cent) component of the $^3D_1$ state (*D*: *L* = 2).

## SCATTERING OF NUCLEONS

Concentrate on scattering in an *L* = 0 state only, further formalism just gets too complicated. For definiteness Look at the scattering in the $^3S_1$ channel, and the $^1S_0$ one. (These are also called the triplet and singlet channels.)

## NUCLEAR FORCES

Having learnt this much about nuclei, what can we say about the nuclear force, the attraction that holds nuclei together? First of all, from Rutherford's old experiments on particle scattering from nuclei, one can learn that the range of these forces is a few fm.

From the fact that nuclei saturate, and are bound, we would then naively build up a picture of a potential that is strongly repulsive at short distances, and shows some mild attraction at a range of 1-2 fm.

Here we assume, that just as the Coulomb force can be derived from a potential that only depends on the size of *r*,

$$V(r) = \frac{q_1 q_2}{4\pi \epsilon_0 r},'$$

the nuclear force depends only on $r$ as well. This is the simplest way to construct a rotationally invariant energy. For particles with spin other possibilities arise as well (e.g., $\hat{S} \cdot$ r).

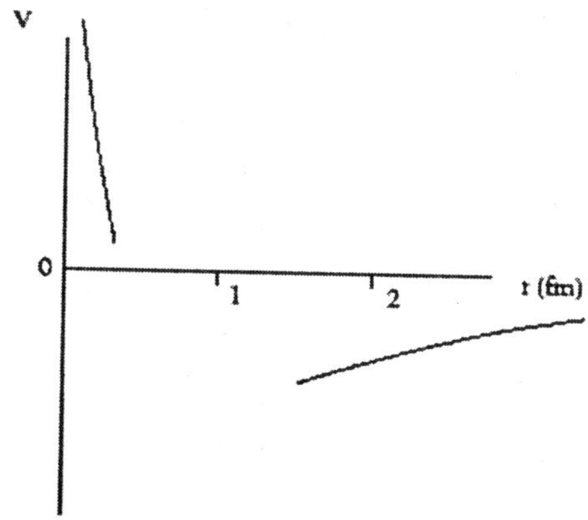


**Fig.** Possible form for the internucleon potential, repulsive at short distances, and attractive at large distances.

Since we have taken the force to connect pairs of particles, we can just study the interaction of two nucleons, by looking both at the bound states (there is only one), and at scattering, where we study how a nucleon gets deflected when it scatters of another nucleon. Let us first look at the deuteron, the bound state of a proton and a neutron. The quantum numbers of its ground state are $J\pi = 1^+$, $I = 0$. A little bit of additional analysis shows that this is a state with $S = 1$, and $L = 0$ or 2. Naively one would expect a lowest state $S = 0$, $L = 0$ (which must have $I =$ for symmetry reasons not discussed here).

We conclude the following:

1. The nuclear force in the $S$ waves is attractive.
2. Nuclear binding is caused by the tensor force.
3. The nuclear force is isospin symmetric (i.e., it is independent of the direction of isospin).

## Chapter 3

# Nuclear Masses

1. Each nucleus has a (positive) charge $Ze$, and integer number times the elementary charge $e$. This follows from the fact that atoms are neutral!
2. Nuclei of identical charge come in different masses, all approximate multiples of the "nucleon mass". (Nucleon is the generic term for a neutron or proton, which have almost the same mass, $m_p$ = 938.272MeV/$c^2$, $m_n$ = MeV/$c^2$.) Masses can easily be determined by analysing nuclei in a *mass spectrograph* which can be used to determine the relation between the charge $Z$ (the number of protons, we believe) vs. the mass.

Nuclei of identical charge (chemical type) but different mass are called isotopes. Nuclei of approximately the same mass, but different chemical type, are called isobars.

### MASS SPECTROGRAPH

A mass spectrograph is a combination of a bending magnet, and an electrostatic device (to be completed).

### INTERPRETATION

We conclude that the nucleus of mass $m \approx Am_N$ contains $Z$ positively charged nucleons (protons) and $N = A - Z$ neutral nucleons (neutrons). These particles are bound together by the "nuclear force", which changes the mass below that of free particles. We write $^A$El for an element of chemical type $^A$El, which determines $Z$, containing $A$ nucleons.

## DEEPER ANALYSIS OF NUCLEAR MASSES

To analyse the masses even better we use the atomic mass unit (amu), which is 1/12th of the mass of the neutral carbon atom,

$$1 \text{ amu} = \frac{1}{12} m_{^{12}\text{C}}.$$

This can easily be converted to SI units by some chemistry. One mole of $^{12}C$ weighs 0.012 kg, and contains Avogadro's number particles, thus

$$1 \text{ amu} = \frac{0.001}{N_A} \text{kg} = 1.66054 \times 10^{-27} \text{ kg} = 931.494 \text{MeV}/c^2.$$

The quantity of most interest in understanding the mass is the binding energy, defined for a neutral atom as the difference between the mass of a nucleus and the mass of its constituents,

$$B(A, Z) = ZM_H c^2 + (A - Z)M_n c^2 - M(A, Z)c^2.$$

With this choice a system is bound when $B > 0$, when the mass of the nucleus is lower than the mass of its constituents. Let us first look at this quantity per nucleon as a function of $A$.

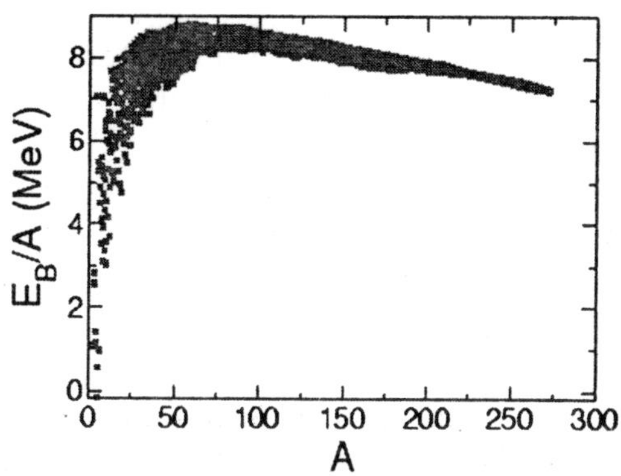


**Fig.** *B*/*A* versus *A*

This seems to show that to a reasonable degree of approximation the mass is a function of *A* alone, and furthermore, that it approaches a constant. This is called nuclear saturation. This agrees with experiment, which suggests that the radius of a nucleus scales with the 1/3rd power of $A$,

$$R_{\text{RMS}} \approx 1.1A^{1/3} \text{ fm}.$$

This is consistent with the saturation hypothesis made by Gamov in the 30's:

As A increase the volume per nucleon remains constant.

For a spherical nucleus of radius $R$ we get the condition

$$\frac{4}{3}\pi R^3 = AV_1,$$

or

$$R = (\frac{V_1 3}{4\pi})^{1/3} A^{1/3}.$$

From which we conclude that

$$V_1 = 5.5 \text{ fm}$$

## NUCLEAR MASS FORMULA

There is more structure than just a simple linear dependence on $A$. A naive analysis suggests that the following terms should play a rôle:

1. Bulk energy: This is the term studied above, and saturation implies that the energy is proportional to $B_{\text{bulk}} = aA$.
2. Surface energy: Nucleons at the surface of the nuclear sphere have less neighbours, and should feel less attraction. Since the surface area goes with $R^2$, we find $B_{\text{surface}} = -\beta A$.
3. Pauli or symmetry energy: nucleons are fermions (will be discussed later). That means that they cannot occupy the same states, thus reducing the binding. This is found to be proportional to $B_{\text{symm}} = -\gamma(N/2 - Z/2)^2/A^2$.
4. Coulomb energy: protons are charges and they repel. The average distance between is related to the radius

of the nucleus, the number of interaction is roughly $Z^2$ (or Z(Z – 1)). We have to include the term $B_{Coul} = -\in Z^2/A$.

Taking all this together we fit the formula

$$B(A, Z) = \alpha A - {}^{\beta}A^{2/3} - {}^{\gamma}(A/2 - Z)^2 A^{-2} - {}^{\varepsilon}Z^2 A^{-1/3}$$

to all know nuclear binding energies with $A \geq 16$ (the formula is not so good for light nuclei). The fit results are given in table.

**Table: Fit of masses to Eq.**

| parameter | value |
|---|---|
| α | 15.36 MeV |
| β | 16.32 MeV |
| γ | 90.45 MeV |
| ε | 0.6928 MeV |

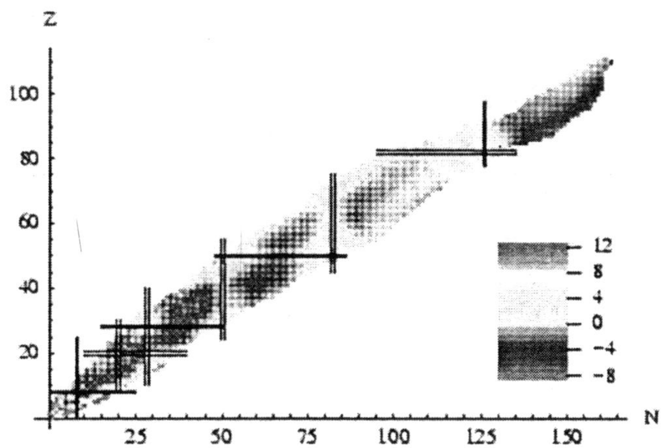


**Fig.** Difference between fitted binding energies and experimental values, as a function of *N* and *Z*.

There remains a certain amount of structure, as well as a strong difference between neighbouring nuclei. This is due to the superfluid nature of nuclear material: nucleons of opposite momenta tend to anti-align their spins, thus gaining energy. The solution is to add a pairing term to the binding energy,

$$B_{\text{pair}} = \begin{cases} \delta A^{-1/2} \text{ for N even, } Z \text{ odd} \\ -\delta A^{-1/2} \text{ for N odd, } Z \text{ even} \end{cases}$$

The results including this term are significantly better, even though all other parameters remain at the same position.Taking all this together we fit the formula

$$B(A, Z) = \alpha A - \beta A^{2/3} - \gamma(A/2 - Z)^2 A^{-2} - \delta B_{\text{pair}}(A, Z) - \varepsilon Z^2 A^{-1/3}$$

**Table: Fit of masses to Eq.**

| **parameter** | **value** |
|---|---|
| α | 15.36 MeV |
| β | 16.32 MeV |
| γ | 90.46 MeV |
| $\delta$ | 11.32 MeV |
| ε | 0.6929 MeV |

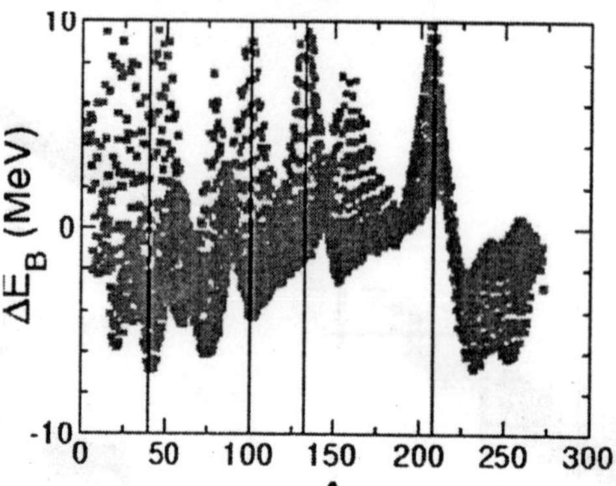


**Fig.** *B*/*A* versus *A*, mass formula subtracted

## STABILITY OF NUCLEI

We have colour coded the nuclei of a given mass $A = N + Z$ by their mass, red for those of lowest mass through to magenta for those of highest mass. We can see that typically the nuclei that are most stable for fixed *A* have more neutrons than protons, more so for large *A* increases than for low *A*. This is the "neutron excess".

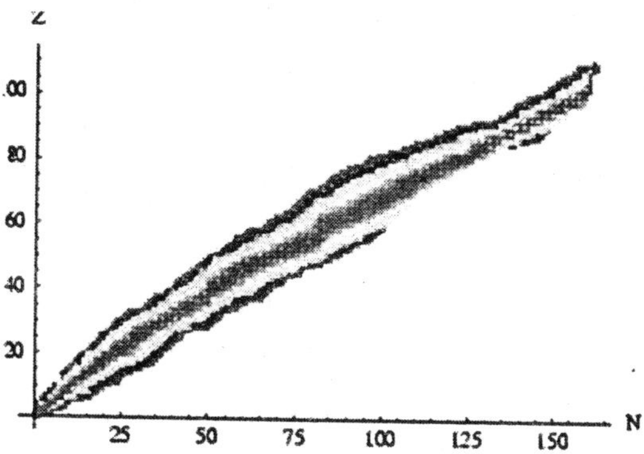


Fig. The valley of stability

## β DECAY

If we look at fixed nucleon number $A$, we can see that the masses vary strongly,

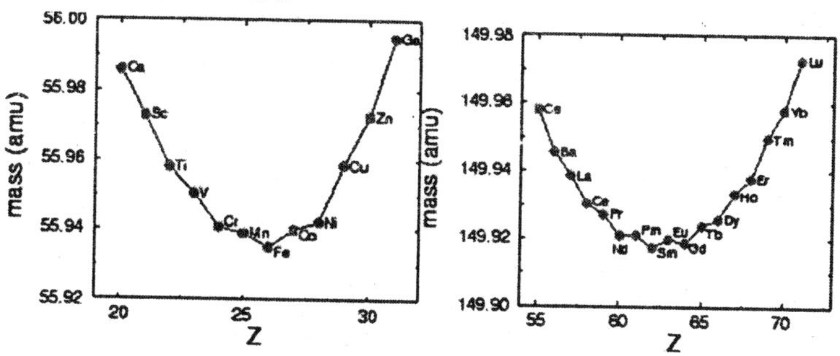


Fig. A cross section through the mass table for fixed $A$. To the left, $A = 56$, and to the right, $A = 150$.

It is known that a free neutron is not a stable particle, it actually decays by emission of an electron and an antineutrino,

$$n \rightarrow p + e^- + \bar{\nu}_e .$$

The reason that this reaction can take place is that it is endothermic, $m_n c^2 > m_p c^2 + m_e c^2$. (Here we assume that the neutrino has no mass.) The degree of allowance of such a reaction is usually expressed in a $Q$ value, the amount of energy released in such a reaction,

$$Q = m_n c^2 - m_p c^2 - m_e c^2 = 939.6 - 938.3 - 0.5 = 0.8 \text{ MeV}.$$

Generically it is found that two reaction may take place, depending on the balance of masses. Either a neutron "$\beta$ decays" as sketched above, or we have the inverse reaction

$$p \rightarrow n + e^+ + \bar{\nu}_e .$$

For historical reason the electron or positron emitted in such a process is called a $\beta$ particle. Thus in $\beta^-$ decay of a nucleus, a nucleus of $Z$ protons and $N$ neutrons turns into one of $Z + 1$ protons and $N - 1$ neutrons. In $\beta^+$ decay the nucleus moves to the left. Using atomic masses, the $Q$ factor is

$$Q\beta^- = M(A, Z)c^2 - M(A, Z + 1)c^2,$$

$$Q\beta^- = M(A, Z)c^2 - M(A, Z - 1)c^2 - 2m_e c^2.$$

The double electron mass contribution in this last equation because the atom looses one electron, as well as emits a positron with has the same mass as the electron.

In similar ways we can study the fact whether reactions where a single nucleon (neutron or proton) is emitted, as well as those where more complicated objects, such as Helium nuclei ($\alpha$ particles) are emitted. But let us note the $Q$ values,

| | |
|---|---|
| neutron emission | $Q = (M(A, Z) - M(A - 1, Z) - m_n)c^2,$ |
| proton emission | $Q = (M(A, Z) - M(A - 1, Z - 1) - M(1, 1))c^2,$ |
| emission | $Q = (M(A, Z) - M(A - 4, Z - 2) - M(4, 2))c^2,$ |
| break up | $Q = (M(A, Z) - M(A - A_1, Z - Z_1) - M(A_1, Z_1))c^2.$ |

Nuclei are quantum systems, and as such must be described by a quantum Hamiltonian. Fortunately nuclear energies are much smaller than masses, so that a description in terms of non-relativistic quantum mechanics is possible. Such a description is not totally trivial since we have to deal with quantum systems containing many particles. Rather then solving such complicated systems, we often resort to models.

## Chapter 4

# Nuclear Models

There are two important classed of nuclear models: single particle and microscopic models, that concentrate on the individual nucleons and their interactions, and collective models, where we just model the nucleus as a collective of nucleons, often a nuclear fluid drop.

Microscopic models need to take into account the Pauli principle, which states that no two nucleons can occupy the same quantum state. This is due to the Fermi-Dirac statistics of spin 1/2 particles, which states that the wave function is antisymmetric under interchange of any two particles

## NUCLEAR SHELL MODEL

The simplest of the single particle models is the nuclear shell model. It is based on the observation that the nuclear mass formula, which describes the nuclear masses quite well on average, fails for certain "magic numbers", i.e., for neutron number $N = 20, 28, 50, 82, 126$ and proton number $Z = 20, 28, 50, 82$. These nuclei are much more strongly bound than the mass formula predicts, especially for the doubly magic cases, i.e., when $N$ and $Z$ are both magic. Further analysis suggests that this is due to a shell structure, as has been seen in atomic physics.

## MECHANISM THAT CAUSES SHELL STRUCTURE

In atoms it is the Coulomb force of the heavy nucleus that forces the electrons to occupy certain orbitals. This can be seen as an external agent. In nuclei no such external force exits, so we have to find a different mechanism.

The solution, and the reason the idea of shell structure in nuclei is such a counter-intuitive notion, is both elegant and simple. Consider a single nucleon in a nucleus. Within this nuclear fluid we can consider the interactions of each of the nucleons with the one we have singled out. All of these nucleons move rather quickly through this fluid, leading to the fact that our nucleons only sees the average effects of the attraction of all the other ones. This leads to us replacing, to first approximation, this effect by an average nuclear potential.

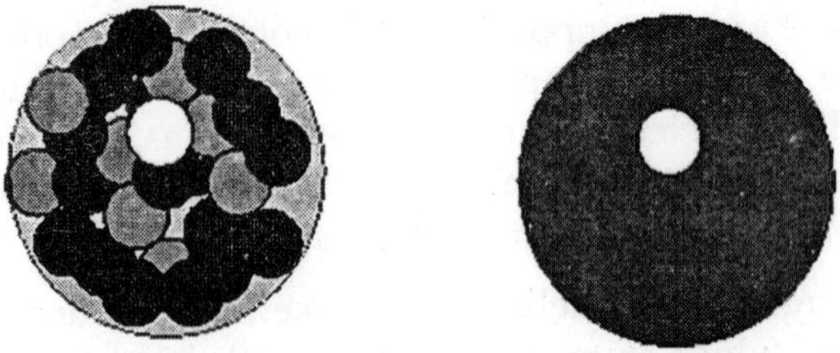

**Fig:** A sketch of the averaging approximation

Thus the idea is that the shell structure is caused by the average field of all the other nucleons, a very elegant but rather surprising notion!

## MODELING THE SHELL STRUCTURE

Whereas in atomic physics we solve the Coulomb force problem to get the shell structure, we expect that in nuclei the potential is more attractive in the centre, where the density is highest, and less attractive near the surface. There is no reason why the attraction should diverge anywhere, and we expect the potential to be finite everywhere. One potential that satisfies these criteria, and can be solved analytically, is the Harmonic oscillator potential. Let us use that as a first model, and solve

$$\frac{\hbar}{2m}\Delta\psi(r)+\frac{1}{2}m\omega^2r^2(r)=E\psi(r).$$

The easiest way to solve this equation is to realise that, since $r^2 = x^2 + y^2 + z^2$, the Hamiltonian is actually a sum of an

$x$, $y$ and $z$ harmonic oscillator, and the eigenvalues are the sum of those three oscillators,

$$E_{nxnynz} = (n_x + n_y + n_z + 3/2)\hbar\omega .$$

The great disadvantage of this form is that it ignores the rotational invariance of the potential. If we separate the Schrodinger equation in radial coordinates as

$$R_{nl}(r)Y_{LM}(\theta, \phi)$$

with $Y$ the spherical harmonics, we find

$$-\frac{\hbar^2}{2m}\frac{1}{r^2}\frac{\partial}{\partial r}\left( r^2\frac{\partial}{\partial r}R(r)\right) + \frac{\hbar^2 L(L+1)}{r^2} R(r) + mr^2R(r) =$$

$E_{nL}R(r)$.

In this case it can be shown that

$$E_{nL} = (2n + L + 3/2)\hbar\omega ,$$

and $L$ is the orbital angular momentum of the state. We use the standard, so-called spectroscopic, notation of *s, p, d, f, g, h, i, j*,... for $L = 0, 1, 2,$....

The real potentials state of the same number of quanta but different $L$ are no longer degenerate, but there are groups of shells with big energy gaps between them. This cannot predict the magic numbers beyond 20, and we need to find a different mechanism. There is one already known for atoms, which is the so-called spin orbit splitting. This means that the degeneracy in the total angular momentum ($j = L\pm1/2$) is lifted by an energy term that splits the aligned from anti-aligned case. The gaps between the groups of shells are in reality much larger than the spacing within one shell, making the binding-energy of a closed-shell nucleus much lower than that of its neighbours.

## EVIDENCE FOR SHELL STRUCTURE

Evidence for the shell structure can be seen in two ways:

1- By looking at nuclear reactions that add a nucleon or remove a nucleon from a closed shell nucleus. The most sensitive of these are electron knockout reactions, where an electron comes in and an electron and a proton or neutron escapes, usually denoted as $(e, e'p)$ $(e, e'n)$ reactions. In those we see clear evidence of peaks at the single particle energies.

2- By looking at nuclei one particle or one hole away from a doubly magic nucleus. As an example look at the nuclei around 208Pb.

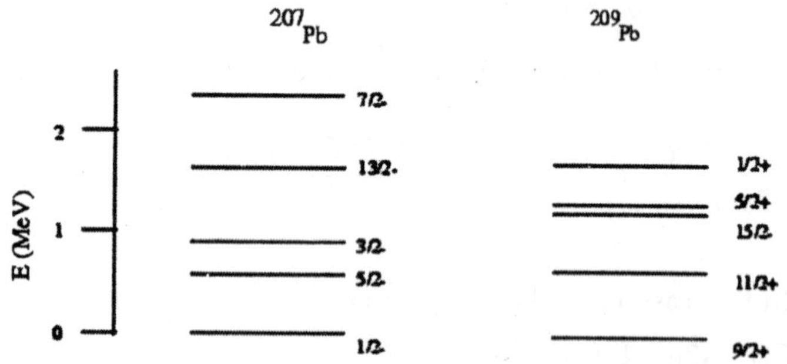


**Fig.** The spectra for the one-neutron hole nucleus $^{207}$Pb and the one-particle nucleus $^{209}$Pb.

In order to understand we need to think a little about the shell structure, as sketched very schematically. The one neutron-hole nucleus corresponds to taking away a single neutron from the 50-82 shell, and the one neutron particle state to adding a a neutron above the $N = 128$ shell closure. We can also understand more clearly why a closed shell nucleus has very few low-energy excited quantum states, since we would have to create a hole below the closed shell, and promote the nucleon in that shell to an open state above the closure. This requires an energy that equals the gap in the single-particle energy.

## COLLECTIVE MODELS

Another, and actually older, way to look at nuclei is as a drop of "quantum fluid". This ignores the fact that a nucleus is made up of protons and neutrons, and explains the structure of nuclei in terms of a continuous system, just as we normally ignore the individual particles that make up a fluid.

## LIQUID DROP MODEL AND MASS FORMULA

Now we have some basic information about the liquid drop model, let us try to reinterpret the mass formula in terms of this model; especially as those of a spherical drop of liquid.

As a prime example consider the Coulomb energy. The general energy associated with a charge distribution is

$$E_{\text{Coulomb}} = \frac{1}{2}\int \frac{\rho(r_1)\rho(r_2)}{4\pi \in_0 r_{12}} d^3r_1 d^3r_2,$$

where the charge distribution is the smeared out charge of the protons,

$$4\pi \int_0^R \rho(r) r^2 dr = Ze$$

If we take the charge to be homogeneously distributed $\rho = Ze/(4/3\pi R^3)$, then

$$E_{\text{Coulomb}} = \frac{(2\pi)(4\pi)}{4\pi \in_0} p^2 \int_0^R \int_{-1}^{1} \frac{r_1^2 dr_2 d\cos\theta r_2^2 dr_2}{(r_1^2 + r_2^2 - 2r_1 r_2 \cos\theta)^{1/2}}$$

$$= \frac{\pi\rho^2}{2\in_0} \int_0^R \int_0^R \frac{(r_1 + r_2) - |r_1 + r_2|}{r_1 r_2} r_1^2 dr_1\, r_2^2 dr_2$$

$$= \frac{\pi\rho^2}{2\in_0}\left[2\int_0^R x^2 dx \int_0^R y dy - \int_0^R \int_0^R |r_1 - r_2| r_1 r_2 dr_1 dr_2\right]$$

$$= \frac{\pi\rho^2}{2\in_0}\frac{R^5}{3} - R^5 15$$

$$= \frac{\pi Z^2 9e^2}{(4\pi)^2 R^6 2\in_0}\frac{4R^5}{15} = \frac{e^2}{4\pi \in_0 \frac{3}{10}}\frac{Z^2}{R}$$

## EQUILIBRIUM SHAPE AND DEFORMATION

Once we picture a nucleus as a fluid, we can ask question about its equilibrium shape. From experimental data we know that near closed shells nuclei are spherical, i.e., the equilibrium shape is a sphere. When both the proton and neutron number differ appreciably from the magic numbers, the ground state is often found to be axially deformed, either prolate (cigar like) or oblate (like a pancake).

A useful analysis to perform is to see what happens when we deform a nucleus slightly, turning it into an ellipsoid, with one axis slightly longer than the others, keeping a constant volume:

$$a = R(1 + \varepsilon), \qquad b = (R(1 + \varepsilon)^{-1/2}.$$

The volume is $\frac{4}{3}\pi\, ab^2$, and is indeed constant. The surface area of an ellipsoid is more complicated, and we find

$$S = 2\pi\left[b^2 + ab,\frac{\arcsin e}{e}\right]$$

where the eccentricity $e$ is defined as

$$e = [(1 - b^2/a^2]^{1/2}.$$

For small deformation we find a much simpler result,

$$S = 4\pi R^2\left[1 + \frac{2}{5}\epsilon^2\right],$$

and the surface area thus increases for both elongations and contractions. Thus the surface energy increases by the same factor. There is one competing term, however, since the Coulomb energy also changes, the Coulomb energy goes down, since the particles are further apart,

$$E_{\text{Coulomb}} \to E_{\text{Coulomb}}\left(1 - \frac{\epsilon^2}{5}\right)$$

We thus find a change in energy of

$$DeltaE = \epsilon^2\left[\frac{2}{5}\beta A^{2/3} - \frac{1}{5}\epsilon z^2 A^{-1/3}\right]$$

The spherical shape is stable if $\Delta E > 0$.

Since it is found that the nuclear fluid is to very good approximation incompressible, the dynamical excitations are those where the shape of the nucleus fluctuates, keeping the volume constant, as well as those where the nucleus rotates without changing its intrinsic shape.

## COLLECTIVE VIBRATIONS

Let us first look at collective vibrations, and for simplicity only at those of a spherical fluid drop. We can think of a large number of shapes; a complete set can be found by parametrising the surface as $r = \sum L, M\ a_{LM} Y_{LM}(\theta, \phi)$, where $Y_{LM}$ are the spherical harmonics and describe the multipolarity (angular momentum) of the surface. The effects of monopole ($L$ = 0), dipole ($L$ = 1), quadrupole ($L$ = 2) and octupole ($L$ = 3) modes. Let us investigate these modes in turn, in the harmonic limit, where we look at small vibrations (small $a_{LM}$) only.

## MONOPOLE

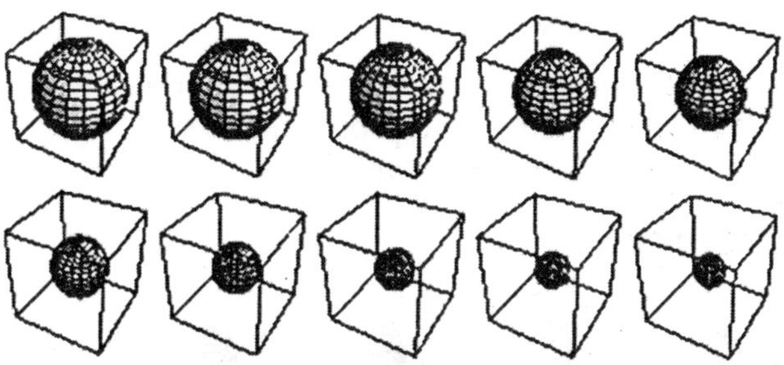

**Fig.** Monopole fluctuations of a liquid drop

The monopole mode, is the one where the size of the nuclear fluid oscillates, i.e., where the nucleus gets compressed. Experimentally one finds that the lowest excitation of this type, which in even-even nuclei carries the quantum number $J\pi = 0^+$, occurs at an energy of roughly

$$E_0 \approx 80A^{-1/3}\text{MeV}$$

above the ground state. Compared to ordinary nuclear modes, which have energies of a few MeV, these are indeed high energy modes ( 15 MeV for $A$ = 216), showing the incompressibility of the nuclear fluid.

**DIPOLE**

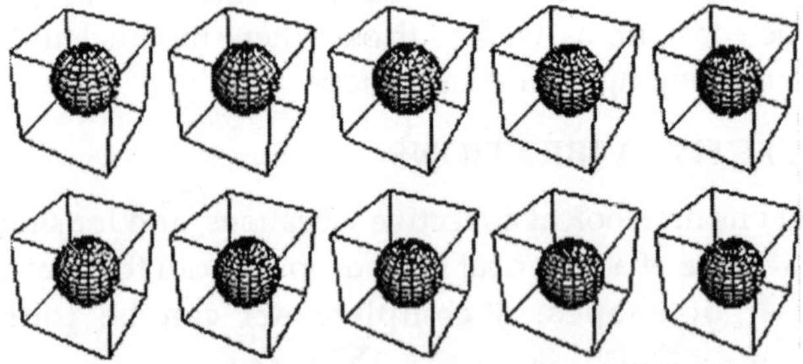

**Fig.** Dipole fluctuations of a liquid drop

The dipole mode, by itself is not very interesting: it corresponds to an overall translation of the centre of the nuclear fluid. One can, however, imagine a two-fluid model where a proton and neutron fluid oscillate against each other. This is a collective isovector ($I$ = 1) mode. It has quantum numbers $J\pi = 1^-$, occurs at an energy of roughly

$$E_0 \approx 77A^{-1/3}\text{MeV}$$

above the ground state, close to the monopole resonance. It shows that the neutron and proton fluids stick together quite strongly, and are hard to separate.

**QUADRUPOLE**

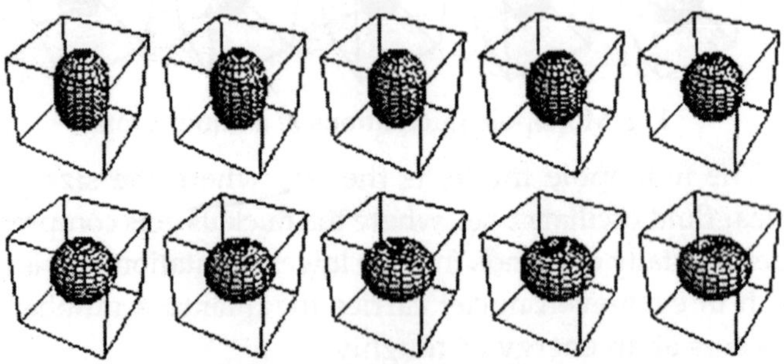

**Fig.** Quadrupole fluctuations of a liquid drop

Quadrupole modes, are the dominant vibrational feature in almost all nuclei. The very special properties of the lower multipolarities mean that these are the first modes available for low-energy excitations in nuclei. In almost all even-even nuclei we find a low-lying state (at excitation energy of less than 1 - 2MeV), which carries the quantum numbers $J\pi = 2^+$, and near closed shells we can often distinguish the second harmonic states as well (three states with quantum numbers $J\pi = 0^+, 2^+, 4^+$).

## OCTUPOLE

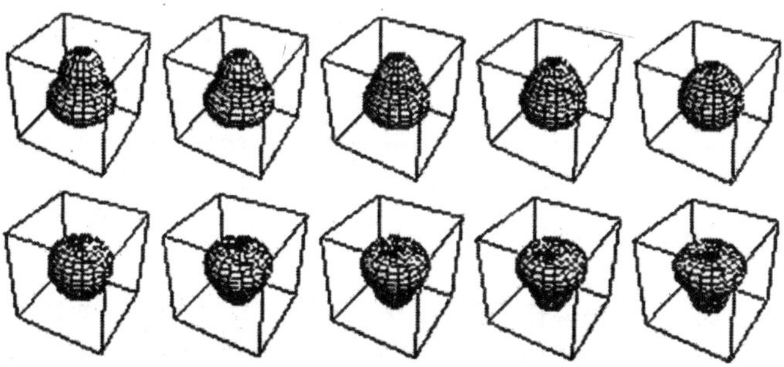

**Fig.** Octupole fluctuations of a liquid drop

Octupole modes, with $J = 3$. can be seen in many nuclei. In nuclei where shell-structure makes quadrupole modes occur at very high energies, such as doubly magic nuclei, the octupole state is often the lowest excited state.

## COLLECTIVE ROTATIONS

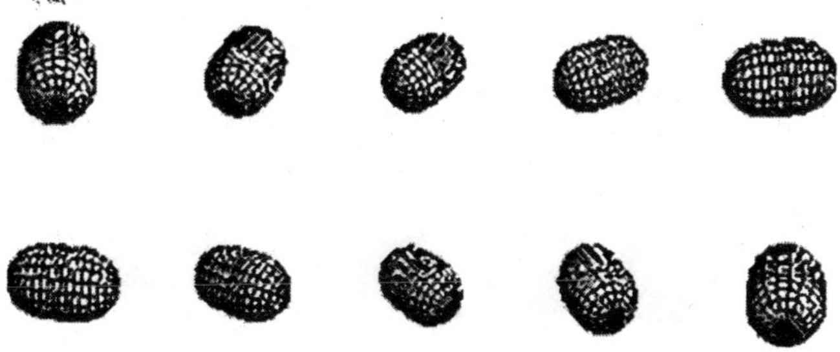

**Fig.** Collective rotation of an axially deformed liquid drop

Once we have created a nucleus with axial deformation, i.e., a nucleus with ellipsoidal shape, but still axial symmetry about one axis, we can rotate the fluid around one of the non-symmetry axes to generate excitations. We cannot do it around a symmtery axis, since the resulting state would just be the same quantum state as we started with, and therefore the energy cannot change. A rotated state around a non-symmetry axis is a different quantum state, and therefore we can overlay many of these states, especially with constant rotational velocity. This is almost like the rotation of a dumbbell, and we can predict the classical spectrum to be of the form

$$H = \frac{1}{2L} J^2,$$

where J is the classical angular momentum. We predict a quantum mechanical spectrum of the form

$$E_{rot}(J) = \frac{\hbar^2}{2L} J(J + 1),$$

where $J$ is now the angular momentum quantum number. Naively we expect the spectrum to be more compressed (the moment of inertial is larger) the more elongated the nucleus becomes. It is known that certain structures in nuclei indeed describe well deformed nuclei, up to super and hyper deformed (axis ratio from 1: 1.2 to 1: 2).

## FISSION

Once we have started to look at the liquid drop model, we can try to ask the question what it predicts for fission, where one can use the liquid drop model to good effect. We are studying how a nuclear fluid drop separates into two smaller ones, either about the same size, or very different in size.

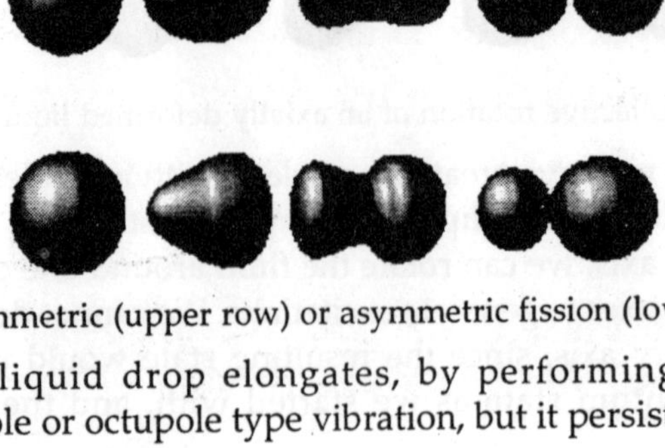

**Fig.** Symmetric (upper row) or asymmetric fission (lower row)

The liquid drop elongates, by performing either a quadrupole or octupole type vibration, but it persists until the nucleus falls apart into two pieces. Since the equilibrium shape must be stable against small fluctuations, we find that the energy must go up near the spherical form.

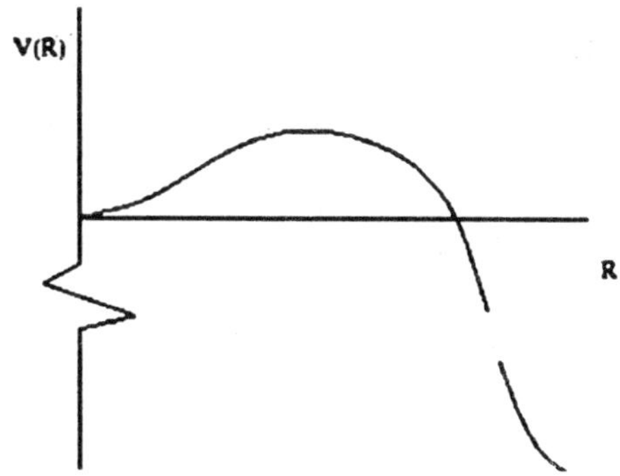


**Fig.** Potential energy for fission

We sketch the energy - which is really the potential energy - for separation into two fragments, *R* is the fragment distance. As with any of such processes we can either consider classical fission decays for energy above the fission barrier, or quantum mechanical tunneling for energies below the barrier. The method used in fission bombs is to use the former, by hitting a $^{235}U$ nucleus with a slow neutron a state with energy above the barrier is formed, which fissions fast. The fission products are unstable, and emit additional neutrons, which can give rise to a chain reaction.

The mass formula can be used to give an indication what is going on; Let us look at at the symmetric fusion of a nucleus. In that case the $Q$ value is

$$Q = M(A, Z) - 2M(A/2, Z/2)$$

Please evaluate this for $^{236}U$ (92 protons). The mass formula fails in prediciting the asymmetry of fission, the splitting process is much more likely to go into two unequal fragments.

## BARRIER PENETRATION

In order to understand quantum mechanical tunneling in fission it makes sense to look at the simplest fission process: the emission of a He nucleus, so called α radiation.

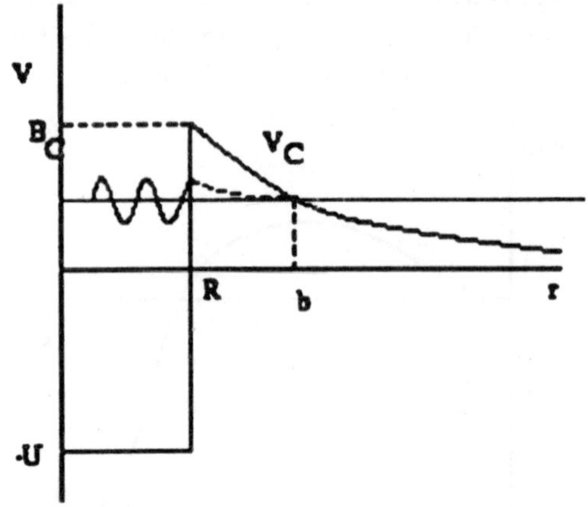


**Fig.** The potential energy for alpha decay

Suppose there exists an particle inside a nucleus at an (unbound) energy > 0. This must be tunneling. In the sketch above we have once again shown the nuclear binding potential as a square well, but we have included the Coulomb tail,

$$V_{\text{Coulomb}}(r) = \frac{(Z-2)2e^2}{4\pi \epsilon_0 r}.$$

The hight of the barrier is exactly the coulomb potential at the boundary, which is the nuclear readius, $R_C = 1.2A^{1.3}$ fm, and thus $B_C = 2.4(Z - 2)A^{-1/3}$. The decay probablility across a barrier can be given by the simple integral expression $P = e^{-2}$, with

$$\gamma = \frac{(2\mu\alpha)^{1/2}}{\hbar} \int_{RC}^{b} \left[V(r) - E^{\alpha}\right]^{1/2} dr$$

$$= \frac{(2\mu\alpha)^{1/2}}{\hbar} \int_{RC}^{b} \left[\frac{2(Z-2)}{4\pi\varepsilon_0 r} - E\alpha\right]^{1/2} dr$$

$$= \frac{2(Z-2)e^2}{2\pi\varepsilon_0 \hbar \upsilon} \left[\arccos(E^{\alpha}/B_C) - (E^{\alpha}/B_C)(1 - E^{\alpha}/B_C)\right],$$

(here $\upsilon$ is the velocity associated with $E^{\alpha}$). In the limit that $B_C \gg E$ we find

$$P = \exp - \left[ \frac{2(Z-2)e^2}{2\varepsilon_0 \hbar \upsilon} \right].$$

This shows how sensitive the probability is to $Z$ and $v$!

## NUCLEAR ENERGY LEVELS

The nucleus, like the atom, has discrete energy levels whose location and properties are governed by the rules of quantum mechanics. The locations of the excited states differ for each nucleus. The excitation energy, $E\ x$, depends on the internal structure of each nucleus. Each excited state is characterized by quantum numbers that describe its angular momentum, parity, and isospin.

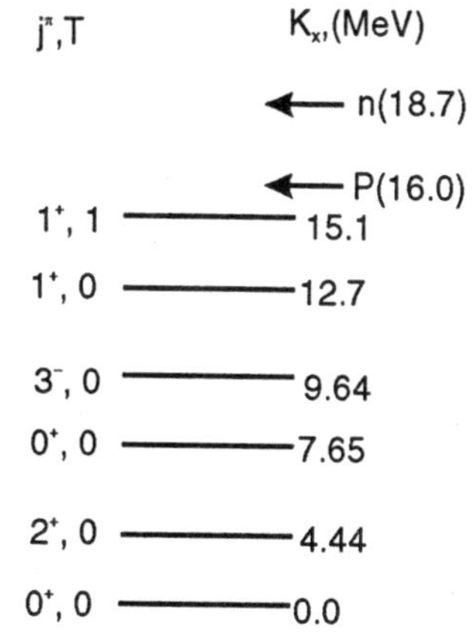


Energy level diagram of some of the excited states of the $^{12}C$ nucleus. The angular momentum ($J$), parity ($P$), and isospin ($T$) quantum numbers of the states are indicated on the left using the notation $J^P$, $T$. The separation energies for a proton and a neutron are indicated by $p$ and $n$ respectively at the top of the diagram.

The angular momentum quantum number, $J$, is the integer or half-integer that is the measure of the total angular momentum of the energy state in units of $\hbar$ (Planck's constant $h$ divided $2\pi$)

$$\text{angular momentum} = J\hbar .$$

The parity, $P$, of a nuclear energy level is a statement about what the nuclear structure of the state would look like if the spatial coordinates of all the nucleons were reversed: $P = +$ means the reversed state would look the same as the original; $P = -$ means the reversed state differs from the original. The isospin (projection) quantum number, $T$, is an integer or half-integer that measures a property that results if neutron and proton coordinates were interchanged. These quantum numbers for each excited state in the notation $J$, $P$, $T$. These quantum numbers are results of the basic symmetries of the underlying force law that governs the binding of nucleons in a nucleus. They determine how an excited state will decay into another state in the same nucleus (gamma decay) or into a specific state in a different nucleus (beta or alpha decay).

Analyzing the interactions among many nucleons to calculate the energy levels and their properties is a complicated mathematical task. Instead, nuclear scientists have developed several nuclear models that simplified the description the nucleus and the mathematical calculations. These simpler models still preserve the main features of nuclear structure.

## The Shell Model

One such model is the Shell Model, which accounts for many features of the nuclear energy levels. According to this model, the motion of each nucleon is governed by the average attractive force of all the other nucleons. The resulting orbits form "shells," just as the orbits of electrons in atoms do. As nucleons are added to the nucleus, they drop into the lowest-energy shells permitted by the Pauli Principle which requires that each nucleon have a unique set of quantum numbers to describe its motion

When a shell is full (that is, when the nucleons have used up all of the possible sets of quantum number assignments), a nucleus of unusual stability forms. This concept is similar to that found in an atom where a filled set of electron quantum numbers results in an atom with unusual stability–an inert gas. When all the protons or neutrons in a nucleus are in filled shells, the number of protons or neutrons is called a "magic number." Some of the magic numbers are 2, 8, 20, 28, 50, 82, and 126. $^{116}Sn$ has a magic number of protons (50) and $^{54}Fe$ has a magic number of neutrons (28). Nuclei, $^{40}Ca$ and $^{208}Pb$, have magic numbers of both protons and neutrons; these nuclei have exceptional stability and are called "doubly magic." Magic numbers are indicated on the chart of the nuclides.

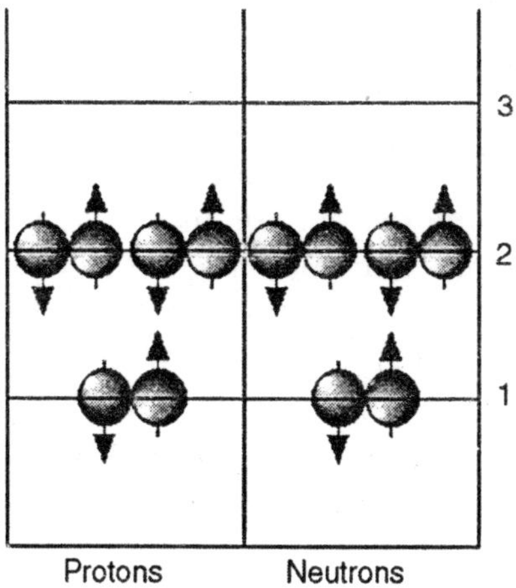


Filled shells have a total angular momentum, *J*, equal to zero. The next added nucleon (a valence nucleon) determines the *J* of the new ground state. When nucleons (singly or in pairs) are excited out of the ground state they change the angular momentum of the nucleus as well as its parity and isospin projection quantum numbers. The shell model describes how much energy is required to move nucleons from one orbit to another and how the quantum numbers change. An energy diagram of the two filled shells of the ground state of $^{12}C$. Promotion of a nucleon or a pair of nucleons to an unfilled shell puts the nucleus into one of the excited states.

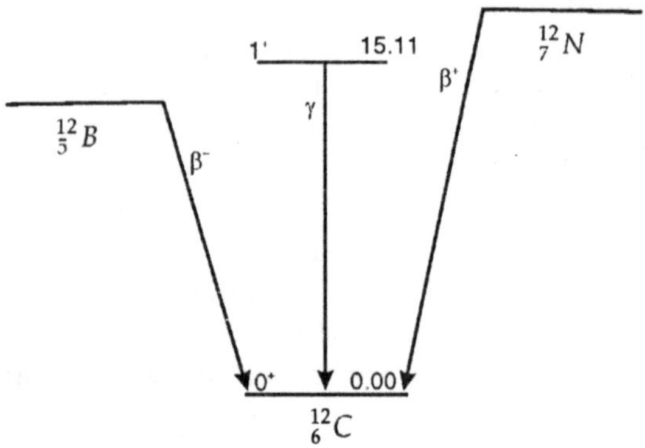


Excited nuclear states decay to more stable states, *i.e.*, more stable nucleon orbitals. Measuring transition rates between nuclear energy levels requires specialized alpha, beta, and gamma detectors and associated electronic circuitry to precisely determine the energy and half-life of the decay. Quantum mechanics and shell-model theory permit nuclear scientists to compute the transition probability (rate of decay) between nuclear states. For nuclei whose structure can be described by a small number of valence nucleons outside filled shells, the Shell Model calculations agree very well with measured values of spin and parity assignments and transition probabilities.

An example of transitions between a nucleus's energy levels is shown above. The ground states of $^{12}B$ (5 protons, 7 neutrons) and $^{12}N$ (7 protons, 5 neutrons) are related to each other and to the 15.1 MeV state in 12C. Each has a nucleon in the third energy level, each has the quantum number 1+, 1, and each decays to the ground state of carbon. The boron and nitrogen beta decay emit an $e^-$ and e+ respectively. The 15.1 MeV state in carbon decays by gamma emission. The shell model calculations of these transition probabilities agree quite well with the measured rates even though different decay mechanisms are at work in each case. Nuclear theorists keep refining the shell model to understand the details of nuclear structure and to make that knowledge available for applications in nuclear technology.

## The Collective Model

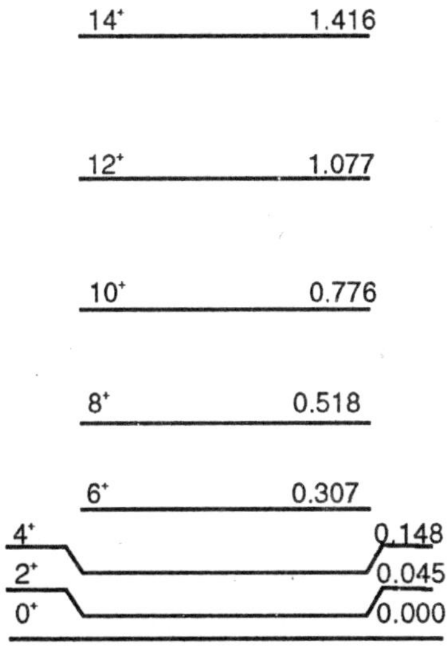


$^{238}_{92}U$

In addition to individual nucleons changing orbits to create excited states of the nucleus as described by the Shell Model, there are nuclear transitions that involve many (if not all) of the nucleons. Since these nucleons are acting together, their properties are called collective, and their transitions are described by a Collective Model of nuclear structure. High-mass nuclei have low-lying excited states that are described as vibrations or rotations of nonspherical nuclei. Many of these collective properties are similar to those of a rotating or vibrating drop of liquid, and in its early development the Collective Model was called the Liquid-Drop Model. The first important application of the Liquid-Drop model was in the analysis of nuclear fission, in which a massive nucleus splits into two lower-mass fragments. The Liquid Drop Model calculates an energy barrier to fission as a sum of the repulsive Coulomb forces between the protons of the nucleus and the attractive surface tension of the skin of the "liquid drop" nucleus. If the barrier is low enough the nucleus might fission spontaneously. For higher barriers, it takes a nuclear reaction to induce fission.

The energy levels of $^{238}U$. The quantum numbers, level spacings, and gamma ray transition probabilities identify these levels as rotational states of a nonspherical nucleus. Nuclei showing collective properties are usually those with many valence nucleons, that is, those with proton or neutron numbers that are far from filled shells. As with the Shell Model, the Çollective Model permits the calculation of spin-parity assignments and transition probabilities that are in good agreement with the measured properties of collective nuclei.

## Measuring Energy Levels

Using accelerators to make nuclear reactions, scientists can create nuclei which have very high angular momentum. Nuclei respond to this rotation, which can be as fast as a hundred billion billion revolutions per second, in a rich and varied way. These nuclei lose some of their excitation energy and almost all of the initial angular momentum by the emission of gamma rays. The gamma ray flash is finished in less than $10^{-9}$ seconds, during which 30 or more gamma rays can be emitted.

A number of preferred pathways in the de-excitation process occur. They relate to favourable arrangements of protons and neutrons and can often be associated with specific symmetries or nuclear shapes. If a sufficient fraction of the decay flows down a particular quantized pathway or band, then the associated structure becomes observable and can be studied in detail. Scientists have recently built arrays of over 100 gamma ray detectors to study the details of some of these rare pathways. Gammasphere was built by groups from Lawrence Berkeley National Laboratory, Argonne National Laboratory, and Oak Ridge National Laboratory. It was used at Berkeley Lab's 88-Inch Cyclotron for two years before being moved to Argonne. During its tenure in Berkeley, many exciting discoveries were made, including:

- *Details of superdeformation in nuclei.* Superdeformation occurs when quantum shell effects help stabilize a football shape (2:1 axis ratio) in certain nuclei. Superdeformed nuclei, prevalent in several regions of the chart of the nuclides, have been found to display some amazing properties.

- *Identical Bands.* Scientists have discovered that sequences of ten or more identical photons are asso ciated with different bands in neighbouring nuclei. This comes as a great surprise; it has long been believed that the gamma-ray emission spectrum for a specific nucleus represents a unique fingerprint. Explaining identical bands is now in the hands of shell model theoreticians.
- *Magnetic Rotation.* Magnetic rotation occurs in nearly spherical nuclei. It is characterized by sequences of gamma-rays reminiscent of collective rotational bands but with a quite different character. Namely, each photon carries off only one (rather than two) units of angular momentum and couples to the magnetic rather than electric properties of the nucleons. This is a new form of quantal rotor that is not fully understood at present.

Fig. One Hemisphere of The Gammasphere Detector Array.

Now Gammasphere has been moved to Argonne National Laboratory where it will be exploring a new range of phenomena. In particular, it will investigate the structure of nuclei far from the line of stability. At Argonne, scientists are studying the fundamental question: why certain combinations of protons and neutrons form stable nuclei while other combinations do not. In other words, it should help us understand what causes the limits on the range of 270 stable isotopes that appear in the chart of the nuclides.

## Chapter 5

# The Fundamental Forces

The fundamental forces are normally divided in four groups, of the four so-called "fundamental" forces. These are often naturally classified with respect to a dimensionless measure of their strength. To set these dimensions we use $\hbar$, $c$ and the mass of the proton, $m_p$. Another important property is their range: the distance to which the interaction can be felt, and the type of quantity they couple to. Let me look a little closer at each of these in turn.

**Table: A summary of the four fundamental forces**

| Force | Range | Strength | Acts on |
|---|---|---|---|
| Gravity | $\infty$ | $G_N \approx 6\ 10^{-39}$ | All particles (mass and energy) |
| Weak Force | $< 10^{-18}$m | $G_F \approx 1\ 10^{-5}$ | Leptons, Hadrons |
| Electromagnetism | $\infty$ | 1/137 | All charged particles |
| Strong Force | $\approx 10^{-15}$m | $g^2 \approx 1$ | Hadrons |

In order to set the scale we need to express everything in a natural set of units. Three scales are provided by $\hbar$ and $c$ and $e$ - actually one usually works in units where these two quantities are 1 in high energy physics. For the scale of mass we use the mass of the proton. In summary (for $e = 1$ we use electron volt as natural unit of energy)

$$\hbar = 6.58 \times 10^{-22} \text{ MeV s}$$

$$\hbar c = 1.97 \times 10^{-13} \text{ MeV}$$

$$m_p = 938 \text{ MeV}/c^2$$

## GRAVITY

The theory of gravity can be looked at in two ways: The old fashioned Newtonian gravity, where the potential is proportional to the rest mass of the particles,

$$V = \frac{G_N m_1 m_2}{r}.$$

We find that $G_N m_p^2 / \hbar c$ is dimensionless, and takes on the value

$$G_N m_p^2 / \hbar c = 5.9046486 \times 10^{-39}.$$

There are two more levels to look at gravity. One of those is Einstein's theory of gravity, which in the low-energy small-mass limit reduces to Newton's theory. This is still a *classical* theory, of a classical gravitational field.

The quantum theory, where we reexpress the field in their quanta has proven to be a very tough stumbling block - When one tries to generalise the approach taken for QED, every expression is infinite, and one needs to define an infinite number of different infinite constants. This is not deemed to be acceptable - i.e., it doesn't define a theory. Such a model is called unrenormalisable. We may return to the problem of quantum gravity later, time permitting.

## ELECTROMAGNETISM

Electro-magnetism, i.e., QED, has been discussed in some detail in the previous chapter. Look there for a discussion. The coupling constant for the theory is

$$\alpha = \frac{e^2}{4\pi\varepsilon_0 \hbar c}$$

## WEAK FORCE

This manifests itself through nuclear $\beta$ decay,

$$\vec{n}p + e^- + \overline{\nu_e}$$

The standard coupling for this theory is called the Fermi coupling, $G_F$, after its discoverer. After the theory was introduced it was learned that there were physical particles that mediate the weak force, the $W^{\pm}$ and the $Z^0$ bosons. These are very heavy particles (their mass is about 80 times the proton mass!), which is why they have such a small range - fluctuations where I need to create that much mass are rare. The $W^{\pm}$ bosons are charged, and the $Z^0$ boson is neutral. The typical $\beta$ decay referred to above is mediated by a $W^-$ boson as can be seen in the Feynman diagram figure. The reason for this choice is that it conserves charge at each point (the charge of a proton and a $W^-$ is zero, the charge of an electron and a neutrino is -1, the same as that of a $W^-$).

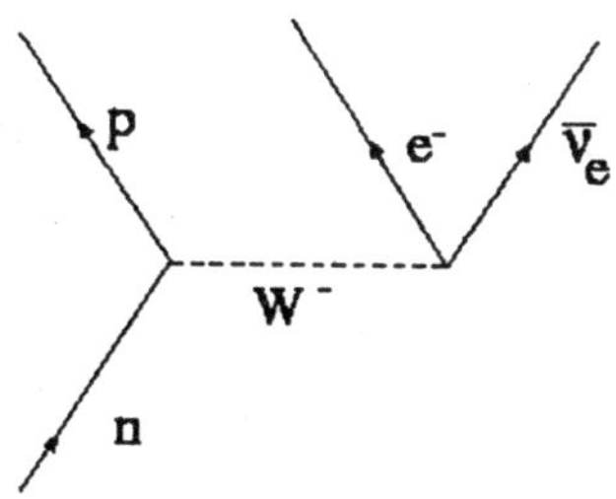


**Fig.** The Feynman diagram for the weak decay of a neutron.

## STRONG FORCE

The strong force is what keeps nuclei together. It is described by a theory called QCD, which described the forces between fermions called quarks that make up the hadrons. These forces are mediated by spin-1 bosons called gluons. Notice that this is a case where a series in powers of the coupling constant does not make a lot of sense, since higher powers have about the same value as lower powers. Such a theory is called non-perturbative.

## SYMMETRIES AND PARTICLE PHYSICS

Symmetries in physics provide a great fascination to us - one of the hang-ups of mankind. We can recognise a symmetry easily, and they provide a great tool to classify shapes and patterns. There is an important area of mathematics called group theory, where one studies the transformations under which an object is symmetric. In order to make this statement seem less abstract, let me look at a simple example, a regular hexagon in a plane. The object is symmetric (i.e., we can't distinguish the new from the old object) under rotations around centre over angles of a multiple of 60°, and under reflection in any of the six axes sketched in the second part of the figure.

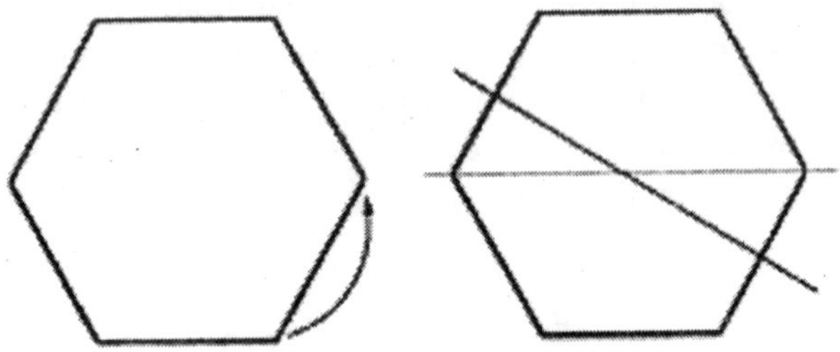

**Fig.** The symmetries of a hexagon

## IMPORTANCE OF SYMMETRIES: NOETHER'S THEOREM

There are important physical consequences of symmetries in physics, especially if the dynamics of a system is invariant under a symmetry transformation.

There is a theorem, due to Emily Noether, one of the most important mathematicians of this century, that states that for any *continuous* symmetry there is a conserved quantity.

So what is a continuous symmetry? Think about something like spherical symmetry - a sphere is invariant under any rotation about its centre, no matter what the rotation angle. The continuity of choice of parameter in a transformation is what makes the set of transformations continuous. Another way of saying the same thing is that the transformation can be arbitrarily close to the unit transformation, i.e., it can do almost nothing at all.

## LORENZ AND POINCARÉ INVARIANCE

One of the most common continuous symmetries of a relativistic theory is Lorentz invariance, i.e., the dynamics is the same in any Lorentz frame. The group of Lorentz transformations can be decomposed into two parts:

- Boosts, where we go from one Lorentz frame to another, i.e., we change the velocity.
- Rotations, where we change the orientation of the coordinate frame.

There is a slightly larger group of symmetries, called the Poincaré group, obtained when we add translations to the set of symmetries – clearly the dynamics doesn't care where we put the orbit of space.

The set of conserved quantities associated with this group is large. Translational and boost invariance implies conservation of four momentum, and rotational invariance implies conservation of angular momentum.

## INTERNAL AND SPACE-TIME SYMMETRIES

Above mentioned angular momentum, the vector product of position and momentum. This is defined in terms of properties of space (or to be more generous, of space-time). But we know that many particles carry the spin of the particle to form the total angular momentum,

$$J = L + S.$$

The invariance of the dynamics is such that J is the conserved quantity, which means that we should not just rotate in ordinary space, but in the abstract "intrinsic space" where S is defined. This is something that will occur several times again, where a symmetry has a combination of a space-time and intrinsic part.

## DISCRETE SYMMETRIES

Let us first look at the key discrete symmetries - parity *P* (space inversion) charge conjugation *C* and time-reversal *T*.

### PARITY *P*

Parity is the transformation where we reflect each point in the origin, $x \rightarrow -x$. This transformation should be familiar to you. Let us think of the one dimensional harmonic oscillator, with Hamiltonian

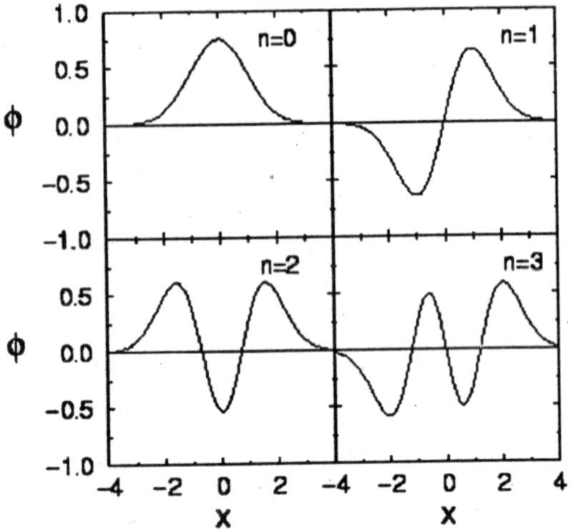


**Fig.** The first four harmonic oscillator wave functions

$$\frac{\hbar^2}{2m}\frac{d}{dx^2}+\frac{1}{2}m\omega^2x^2.$$

The Hamiltonian does not change under the substitution $x \to -x$. The well-known eigenstates to this problem are either even or odd under this transformation and thus have either even or odd parity,

$$P\psi(x, t) = \psi(-x, t) = \pm\psi(x, t),$$

where $P$ is the transformation that take $x \to -x$. For

$$P\psi(x, t) = \psi(x, t)$$

we say that the state has even parity, for the minus sign we speak about negative parity. These are the only two allowed eigenvalues, as can be seen from looking at the probability density $|\psi(x, y)|^2$. Since this must be *invariant*, we find that

$$|\psi(x, y)|^2 = |P\psi(x, y)|^2$$

which shows that the only *real* eigenvalues for $P$ are ±1. One can show that there is a relation between parity and the orbital angular momentum quantum number $L$, $\pi = (-1)^L$, which relates two space-time symmetries.

It is found, however, that parity also has an intrinsic part, which is associated with each type of particle. A photon (γ) has negative parity. This can be understood from the following classical analogy. When we look at Maxwell's equation for the electric field,

$$\nabla E(x,t) = \frac{1}{\varepsilon_0 \rho}(x,t),$$

we find that upon releversal of the coordinates this equation becomes

$$\nabla E(-x,t) = \frac{1}{\varepsilon_0 \rho}(-x,t),$$

The additional minus sign, which originates in the change of sign of $\nabla$ is what gives the electric field and thus the photon its negative intrinsic parity.

We also wish to understand the parity of particles and antiparticles. For fermions (electrons, protons,...) we have the interesting relation $P_f P\bar{f} = -1$, which will come in handy later!

## CHARGE CONJUGATION *C*

The name of this symmetry is somewhat of a misnomer. Originally it stems from QED, where it was found that a set of interacting electrons behaves exactly the same way as a similar set of positrons. So if we change the sign of all charges the dynamics is the same. Actually, the symmetry generalises a little bit, and in general refers to a transformation where we change all particles in their antiparticles.

Once again we find $C^2 = 1$, and the only possible eigenvalues of this symmetry are ±1. An uncharged particle like the photon that is its own antiparticle, must be an eigenstate of the symmetry operation, and it is found that it has eigenvalue –1,

$$C\psi_\gamma = -\psi_\gamma.$$

(Here $\psi_\gamma$ is the wave function of the photon.) This can be shown from Maxwell's equation as before, since $\rho$ chanegs sign under charge conjugation.

For a combination of a particle and an antiparticle, we find that $C_f C \overline{f}$ = - 1 for fermions, and +1 for bosons.

## TIME REVERSAL *T*

On a microscopic scale it is not very apparent whether time runs forward or backwards, the dynamics where we just change the sign of time is equally valid as the original one. This corresponds to flipping the sign of all momenta in a Feynman diagram, so that incoming particles become outgoing particles and vice-versa. This symmetry is slightly nastier, and acts on both space-time and intrinsic quantities such as spin in a complicated way. The space time part is found to be

$$T\,\psi\,(\mathrm{r}, t) = \psi^{\bullet}\,(\mathrm{r}, -t).$$

Combined with its intrinsic part we find that it has eigenvalues $\pm i$ for fermions (electrons, etc.) and ±1 for bosons (photons, etc.).

## THE *CPT* THEOREM

A little thought shows that all three symmetries mentioned above appear very natural - but that is a theorist's argument. The real key test is experiment, not a theorist's nice ideas! In 1956 C.N. Yang and T.D. Lee analysed the experimental evidence for these symmetries. They realised there was good evidence of these symmetries in QED and QCD (the theory of strong interactions). There was no evidence that parity was a symmetry of the weak interactions - which was true, since it was shown soon thereafter that these symmetries are broken, in a beautiful experiment led by "Madame" C.S. Wu.

There is a fairly strong proof that only minimal physical assumptions (locality, causality) that the product of *C*, *P* and

*T* is a good symmetry of any theory. Up to now experiment has not shown any breaking of this product. We would have to rethink a lot of basic physics if this symmetry is not present.

## *CP* VIOLATION

The first experimental confirmation of symmetry breaking was found when studying $\beta^-$ the decay of $^{60}$Co,

$$^{60}Co \rightarrow {}^{60}Ni + e^- + \overline{\nu_e} \, .$$

This nucleus has a ground state with non-zero spin, which can be oriented in a magnetic field.

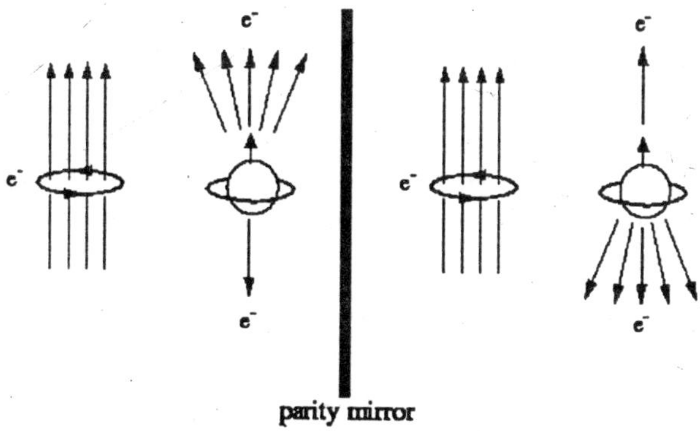


A magnetic field is a pseudo-vector, which means that under parity it goes over into itself B $\rightarrow$ B. So does the spin of the nucleus, and we thus have established that under parity the situation under which the nucleus emits electrons should be invariant. But the direction in which they are emitted changes! Thus any asymmetry between the emission of electrons parallel and anti-parallel to the field implies parity breaking.

# Chapter 6

# Phases of Nuclear Matter

As we know, water ($H_2O$) can exist as ice, liquid, or steam. At atmospheric pressure and below the 0°C freezing point of water we have ice. Between 0°C and 100°C, water is a liquid. Above the boiling point of 100°C, water becomes the gas that we call steam. We also know that we can raise the temperature of water by heating it, that is to say adding energy to it.

However, when water reaches its melting or boiling points, additional heating does not immediately lead to a rise in temperature. Instead, we have to overcome the latent heats of fusion (80 kcal/kg, at the melting point) or vapourization (540 kcal/kg, at the boiling point). During the boiling process, as we add more heat, more of the liquid water turns into steam. Even though heat is being added, the temperature stays at 100ºC. As long as there is still liquid water left, the gas and liquid phase are in coexistence. The temperature cannot rise until all of the liquid is converted to steam. This type of transition between two phases with a latent heat and phase coexistence is called a "first order phase transition."

As we raise the pressure, the boiling temperature of water increases until the critical point at a pressure of 22.1 Mpa (218 times atmospheric pressure) and a temperature of 374°C is reached. There, the phase coexistence stops, and the phase transition becomes continuous or "second order." We can make a diagram, which shows the states of $H_2O$ depending on pressure and temperature.

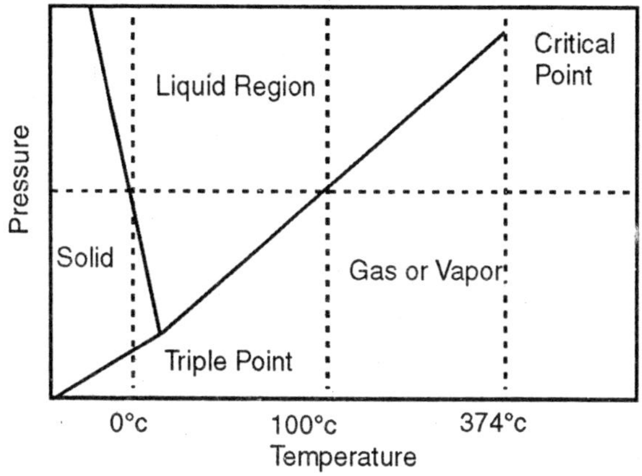


**Fig.** The Phase Diagram for Water.

This diagram indicates that as the pressure increases, ice can turn into water. The diagram can be used to predict the state of $H_2O$ at any temperature and pressure. We call the mathematical relations inferred by the chart the "equation of state" of water.

Just as the state of a collection of atoms or molecules depends on temperature and pressure, we find that the state of a nucleus depends on temperature and on the density of the nucleons. Thus we may ask what is the equation of state for nuclear matter? In their normal states of lowest energy, nuclei show liquid-like characteristics and have a density of 0.17 nucleons/fm$^3$. In more conventional units, this corresponds to $2.7 \times 10^{17}$ kg/m$^3$, or 270 trillion times the density of liquid water.

In a laboratory, the only possible way to heat nuclei to significant temperatures is by colliding them with other nuclei. The temperatures reached during these collisions are astounding. In atomic physics, the electron-volt ($1.6\times10^{-19}$ joules) is used as a convenient unit because it is roughly the energy scale of atomic and chemical processes. Similarly, nuclear scientists use millions of electron-volts or MeV ($1.6\times10^{-13}$ joules) as a convenient energy unit because it is roughly the energy scale of nuclear processes. An average energy of 1 MeV corresponds to a temperature of $1.2 \times 10^{10}$

K. The temperatures we can reach in nuclear collisions range up to 100 MeV and above–more than 200 million times the temperature at the surface of the Sun (~5,500 K)!

As we heat nuclei to a temperature of a few MeV, some of the nuclear "liquid" evapourates. From knowing the general form of the interactions between nucleons, we know that, just like water, the nuclear liquid also has a latent heat of vapourization, and nuclei should undergo a first-order phase transition. This liquid-gas coexistence is also expected to terminate at a critical point, the critical point of nuclear matter. One of the major thrusts of heavy ion research at laboratories such as Michigan State University's National Superconducting Cyclotron Laboratory is to find out if these theoretical expectations are correct. Experiments try to determine at what temperature and density the critical point of nuclear matter is located.

There are major challenges that nuclear physicists face in their efforts to explore the nuclear equation of state and these nuclear phase transitions. We can only establish the hot and dense conditions needed for this process during heavy ion collisions. Thus we do not have the luxury to carefully prepare our sample at a given pressure, temperature and density, as is done when studying the phase diagram of water. Instead we have only a time interval of about $10^{-21}$ seconds during which to conduct our experiment. To complicate things even more, our sample does not stay at a given density and temperature, but expands and cools during our experiment. We also do not have any direct way of measuring the state variables (temperature, pressure, density). We need to determine them from observables such as:

1. The abundance of isotopes,
2. The population of excited nuclear states,
3. The shapes of the energy spectra from nuclear collision remnants
4. The production of particles such as pions.

It is also not obvious that thermal equilibrium can be established during these short time scales. Finally, there is the problem of finite particle number. When studying the phases of water, the sample usually contains very large numbers of molecules. This, again, is a luxury not at our disposal during heavy ion collisions, where the number of nucleons is at best only a few hundred. We then have to figure out what signatures of a phase transition remain when there are only so few elementary constituents present.

Despite these challenges, progress in this field has been significant. We now have deep insight as how to measure the thermodynamics state variables during heavy ion collisions. We are confident that thermal equilibrium can be established, have found evidence for phase coexistence, and are beginning to pin down the critical point of the nuclear liquid-vapour phase diagram. Information about the size of fragments produced when nuclear matter is near its critical point gives essential information on the nuclear equation of state. Recent experiments on nuclear breakup are leading to improved understanding of this important question.

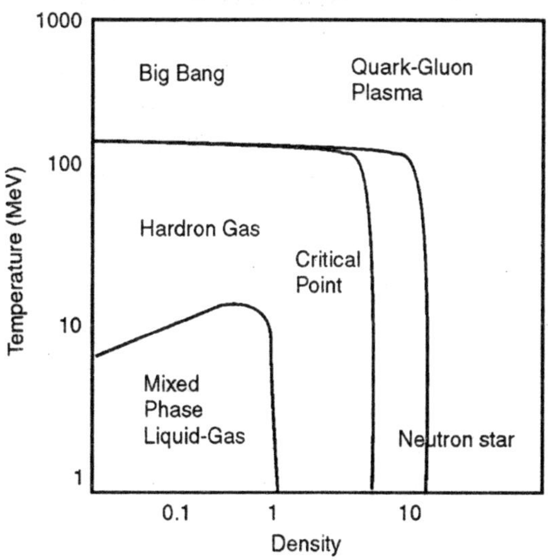


The phase diagram for nuclear matter, as predicted theoretically. The horizontal axis shows the matter density, and the vertical axis shows the temperature. Both axes are shown in logarithmic scale, and the density is given in multiples of normal nuclear matter density. The temperature axis is reversed from the phase diagram of water.

The phase transition between the nuclear liquid and a gas of nucleons is not the only phase transition that heavy ion scientists are studying. At even higher temperatures and densities, nucleons themselves can undergo a phase transition.

We can view each nucleon as a "bag" containing quarks and gluons. These quarks and gluons can move relatively freely inside their own bag, but "bag" theory says that they cannot escape from the bag–they are "confined." For this reason, we have never been able to detect individual free quarks or gluons. However, if we are able to produce an extremely dense gas of hadrons (mainly pions and nucleons), then their bags can overlap. This overlap lets the quarks and gluons from different bags to mix freely and travel across the entire nuclear volume. We call this state a "quark-gluon plasma," in analogy with an atomic plasma in which electrons become unbound from atoms. From theoretical calculations, we also expect the phase transition to a quark-gluon plasma to be of first order, with a phase coexistence region. Major research efforts at BNL (Brookhaven National Laboratory) in New York and at CERN (Conseil Europeen pour la Recherche Nucleaire) in Switzerland are directed toward establishing the conditions for creating this phase transition and observing its signatures. The Relativistic Heavy Ion Collider (RHIC), which has been constructed BNL, has as its main mission the study of this exotic and unique phase transition in the nuclear equation of state. It will produce two counter-circulating beams of gold nuclei, each at speeds extremely close to the speed of light, and collide them.

This is a similar phase diagram to the phase diagram for nuclear matter. A representation of the quarks is shown in the hadron gas area. The mixed phased region shows the existence of both hadrons and quarks and gluons. The quark-gluon plasma region shows the existence of quarks and gluons.

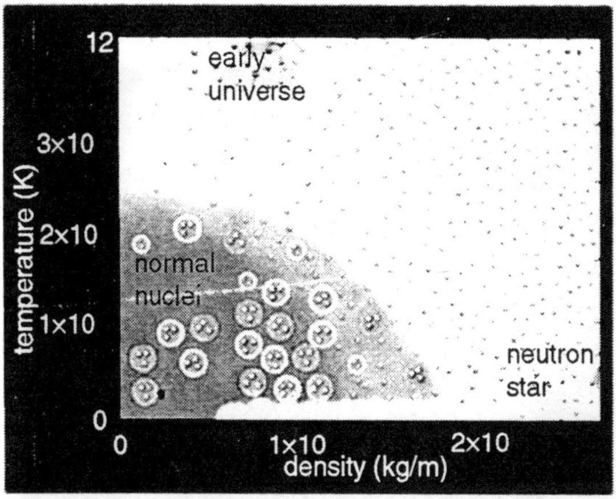


All of the theoretical and experimental challenges listed above are also present when studying the transition to the quark-gluon plasma. In addition, there is another, possibly even more severe obstacle to overcome: the quark-gluon plasma cannot survive longer than a few times $10^{-22}$ seconds. After that, the density and temperature reached during a heavy ion collision fall to values that force quarks and gluons to recombine into hadrons again. The number of hadrons produced in relativistic heavy ion collisions is staggering. For every nucleon that was initially contained in the two gold nuclei, we can expect between 10 and 50 pions to be produced during each RHIC collision. This amounts to many thousands of hadrons emerging from each relativistic heavy ion collision. The essential problem for the nuclear scientist is then to distinguish between those hadrons that were created from the ashes of the quark-gluon plasma and those that were created in the dense hadron gas.

A simulation of a single collision in one of the new detectors (STAR). Each line represents a particle that is tracked and measured in the magnetic field of the detector. The detector measures the thousands of particles produced by the collision. Analysis of the data is like an attempt to "put Humpty-Dumpty together again"–to see if the collision showed any evidence that a quark-gluon plasma had been formed.

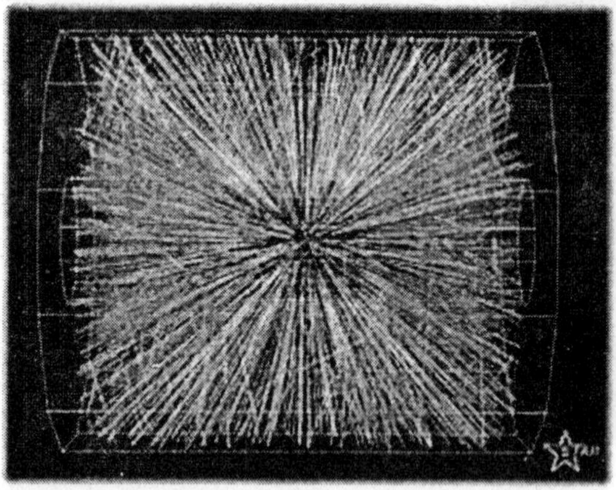

**Fig.** Collision between two gold nuclei

Collision between two gold nuclei in the STAR detector at the RHIC accelerator. Each line represents the path of a particle produced in the collision.

What is the purpose of studying the nuclear matter phase diagram? The answer is that we need this information to understand the early history of our universe, and to understand high-density objects, called "neutron stars" in our present-day universe.

Approximately 73percent of the mass of the visible universe is in the form of hydrogen. Helium makes up about 25percent of the mass, and everything else represents only 2percent. While the abundance of these more massive ("heavy", $A > 4$) elements seems quite low, it is important to remember that most of the atoms in our bodies and Earth are a part of this small portion of the matter of the universe. The low-mass elements, hydrogen and helium, were produced in the hot, dense conditions of the birth of the universe itself. The birth, life, and death of a star is described in terms of nuclear reactions. The chemical elements that make up the matter we observe throughout the universe were created in these reactions.

Approximately 15 billion years ago the universe began as an extremely hot and dense region of radiant energy, the Big Bang. Immediately after its formation, it began to expand and cool. The radiant energy produced quark-antiquarks and electron-positrons, and other particle-antiparticle pairs. However, as the particles and antiparticles collided in the high energy gas, they would annihilate back into electromagnetic energy. As the universe expanded the average energy of the radiation became smaller. Particle creation and annihilation continued until the temperature cooled enough that pair creation became no longer energetically possible.

One of the signatures of the Big Bang that persists today is the long-wavelength radiation that fills the universe. This is radiation left over from the original explosion. The present temperature of this "background" radiation is 2.7 K. (The temperature, $T$, of a gas or plasma and average particle kinetic energy, $E$, are related by the Boltzmann constant, $k = 1.38 \times 10^{-23}$ J/K, in the equation $E = kT$.) The temperature at various stages in the time evolution of the universe from the quark-gluon plasma to the present time.

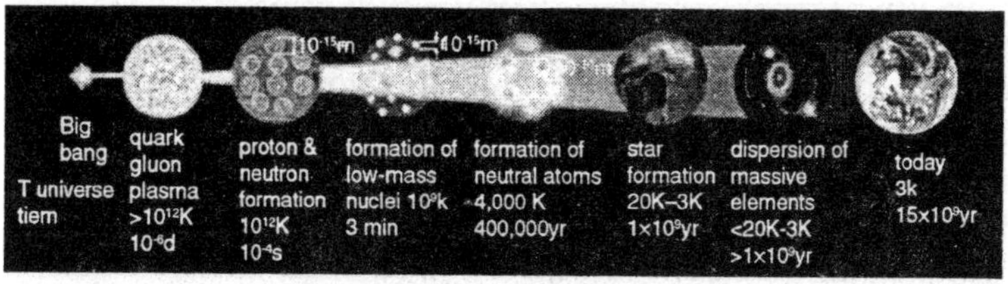


**Fig.** The Evolution of The Universe

At first quarks and electrons had only a fleeting existence as a plasma because the annihilation removed them as fast as they were created. As the universe cooled, the quarks condensed into nucleons. This process was similar to the way steam condenses to liquid droplets as water vapour cools. Further expansion and cooling allowed the neutrons and some of the protons to fuse to helium nuclei. The 73percent hydrogen and 25percent helium abundances that exists throughout the universe today comes from that condensation period during the first three minutes. The 2percent of nuclei more massive than helium present in the universe today were created later in stars.

The nuclear reactions that formed $^4He$ from neutrons and protons were radiative capture reactions. Free neutrons and protons fused to deuterium (d or $^2H$) with the excess energy emitted as a 2.2 MeV gamma ray,

$$n + p \rightarrow d + \gamma.$$

These deuterons could then capture another neutron or free proton to form tritium ($^3H$) or $^3He$,

$$d + n \rightarrow {}^3H + \gamma \text{ and } d + p \rightarrow {}^3He + \gamma.$$

Finally, $^4He$ was produced by the reactions:

$$d+ d \rightarrow {}^4He + \gamma,\ {}^3He + n \rightarrow {}^4He + \gamma \text{ and } {}^3H + p \rightarrow {}^4He + \gamma.$$

Substantial quantities of nuclei more massive than $^4He$ were not made in the Big Bang because the densities and energies of the particles were not great enough to initiate further nuclear reactions.

It took hundreds of thousands of years of further cooling until the average energies of nuclei and electrons were low enough to form stable hydrogen and helium atoms. After about a billion years, clouds of cold atomic hydrogen and helium gas began to be drawn together under the influence of their mutual gravitational forces. The clouds warmed as they contracted to higher densities. When the temperature of the hydrogen gas reached a few million kelvin, nuclear reactions began in the cores of these protostars. Now more massive elements began to be formed in the cores of the very massive stars.

### The Sun

The Sun produces $4 \times 10^{26}$ joules per second of electromagnetic radiation, a fraction of which is intercepted by Earth. The source of this energy is a series of reactions that converts four protons into one helium nucleus plus 26.7 MeV of energy that appears as energy in the reaction products. Since 1 MeV is equivalent to $1.6 \times 10^{-13}$ J there must be

$$\frac{\left(4\times10^{26}\,J/s\right)}{\left(26.7\,MeV/reaction\right)\times\left(1.6\times10^{-13}\,J/MeV\right)} = 0.94\times10^{38}$$

reactions/s

occurring in the sun to maintain its energy flow. The basic reaction chain (86percent of the time) is the fusion sequence:

$^{1}H + {}^{1}H \rightarrow {}^{2}H + e + \alpha,$

$^{2}H + {}^{1}H \rightarrow {}^{3}He + \gamma,$

$^{3}He + {}^{3}He \rightarrow {}^{4}He + {}^{1}H + {}^{1}H.$

These fusion reactions occur only at the centre of the Sun where the high temperature (~$10^7$ K) gives the hydrogen and helium isotopes enough kinetic energy to overcome the long-range repulsive Coulomb force and come within the short-range of the attractive strong nuclear force. The reaction energy slowly percolates to the surface of the Sun where it is radiated mainly in the visible region of the electromagnetic spectrum. Only the neutrinos escape from the Sun without giving up their energy.

A detailed mathematical model of the temperature and density profile of the Sun powered by nuclear reactions also serves as a model of other stars. Since we cannot observe the nuclear reactions directly for confirmation of the nuclear processes, astrophysicists look to the neutrinos produced in the fusion of two protons to form deuterium and in the less common (14percent) branch of the reaction chain where the fusion of $^{3}He$ with $^{4}He$ leads to isotopes of beryllium and boron that emit neutrinos. Massive underground neutrino detectors have found fewer neutrinos than expected from the model calculations. One speculation on the missing neutrinos is that they convert from neutrinos associated with electron processes to those associated with muon or tau processes as they transit from the interior of the Sun to Earth. Such a conversion can only occur if at least one of the neutrino species has a non-zero mass. This is a topic of much current research interest.

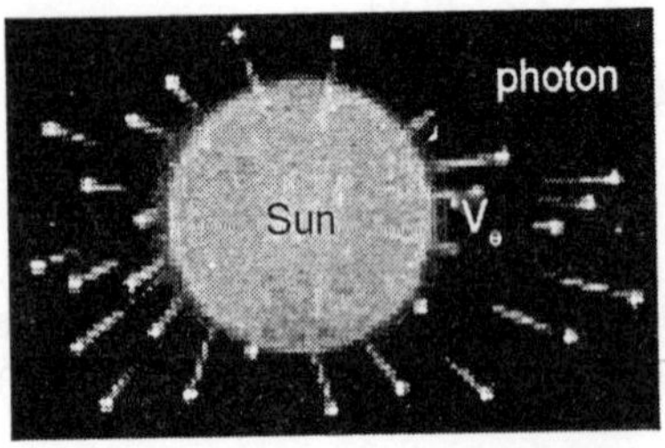


**Fig.** The Solar Energy Reactions Emit Photons and Neutrinos.

### Other Stars

A star the size of the Sun will burn hydrogen into helium until the hydrogen in the core is exhausted. At this point, the core of the star contracts and heats up until the fusion of three $^4$He nuclei into $^{12}$C can begin. Stars in this stage of evolution are known as red giants. Low mass stars such as our Sun will then evolve into a compact object called a white dwarf. All nuclear reactions in a white dwarf have stopped. Higher mass stars have internal temperatures ($10^8$ K) that allow the fusion of carbon with helium to produce oxygen nuclei and excess energy. For very massive stars, the exothermic fusion of low-mass nuclei into successively more massive nuclei can proceed all the way up to nuclei in the iron region (A ~ 60). The temperature (1 keV is equivalent to $1.16 \times 10^7$ K), interior density, and process lifetime that occur in stellar evolution of a star 25 times more massive that the Sun. Note the accelerating time-scale as higher mass nuclei are burned.

Once the core of the star is converted into iron-region nuclei, the star has nearly reached the end of its life. Because the average binding energy per nucleon reaches a maximum at this point, there are no further energy-generating reactions possible and the star collapses because the gravitational force cannot be counter balanced by the high-temperature, high-pressure interior.

As the collapse of the core occurs, the density grows to the point where it becomes energetically favourable for electrons to be captured by protons via the weak interaction, producing neutrons and neutrinos. This process turns the core of the star into neutrons and produces a huge burst of neutrinos. When the collapse reaches nuclear density, the star rebounds explosively, throwing off much of its mass consisting of elements up to iron. This explosive expansion is called a supernova, one of the most spectacular events in astronomy. If the mass of the remnant core is less than two to three times the mass of the Sun, the core will settle down as a compact neutron star with no further nuclear reactions. More massive cores continue to contract due to the intense gravitational force until the size of the core diminishes to a point, a singularity called a black hole. The object is black because the gravitational force is so strong nothing, not even light can escape.

**Table. The Major Stages in The Evolution of a Massive Star.**

| Burning Stage | Temperature (keV) | Density ($kg/m^3$) | Time-scale |
|---|---|---|---|
| Hydrogen | 5 | $5 \times 10^6$ | $7 \times 10^6$ yr |
| Helium | 20 | $7 \times 10^8$ | $5 \times 10^5$ yr |
| Carbon | 80 | $2 \times 10^{11}$ | 600 yr |
| Neon | 150 | $4 \times 10^{12}$ | 1 yr |
| Oxygen | 200 | $10^{13}$ | 6 months |
| Silicon | 350 | $3 \times 10^{13}$ | 1 day |
| Collapse | 600 | $3 \times 10^{15}$ | seconds |
| Bounce | 3000 | $10^{17}$ | milliseconds |
| Explosive | 100-600 | varies | 0.1-10 seconds |

In February of 1987 a supernova in a nearby galaxy was observed in the Southern sky. Underground neutrino detectors saw the neutrinos emitted during the few seconds of the collapse and the birth of either a neutron star or a black hole. The supernova continued to glow for months in the night sky due to the decay of radioactive isotopes that were produced in the explosion. Neutron-capture reactions on iron-region nuclei during the few moments of the explosion produced nuclei more massive than A = 60. A sequence of such reactions can produce elements all the way up to uranium, A = 238.

## SYMMETRIES AND ANTIMATTER

Scientists use symmetry both to solve problems and to search for new understandings of the world around us. In nuclear science the concept of symmetry plays a key role in gaining an understanding of the physical laws governing the behaviour of matter. It provides a shortcut based on geometry for getting at some of Nature's innermost secrets. Because the laws of physics are the same at any time (symmetry in time) and any location (symmetry in position) as another, the laws of conservation of energy and momentum apply.

### SYMMETRY IN ROTATION

Consider the simple idea that when an object is rotated through an angle of 360° it should end in a state no different from its initial state. If we apply this simple symmetry in quantum mechanics, the physics theory of matter and energy at the smallest distances, we find that it imposes the requirement that the angular momentum of rotating objects must be quantized in units of $\hbar$ (Planck's constant, h, divided by $2\pi$). A spinning object, be it a planet, a top, or a nucleus, should only be able to have rotations such that its angular momentum comes out in "chunks" $\hbar$ in size.

Imagine, then the big surprise that swept through the world of physics when it was discovered in the 1930s that this symmetry was "broken" by particles like electrons, protons and neutrons, which were found to have "spin 1/2", or one half an $\hbar$ unit as their "built-in" angular momentum (called "intrinsic angular momentum" or simply "spin"). In nuclear science it has become standard to use $\hbar$ as the measuring stick for angular momentum and to describe the angular momentum of nuclei in units of $\hbar$. Thus, we say that a nucleus has angular momentum 0, or 2, or 7/2, in units of $\hbar$.

One consequence of the half-integer spins of neutrons and protons is that nuclei with an odd number of nucleons must have half-integer angular momentum, while nuclei having an even number of nucleons must have integer angular momentum (in $\hbar$ units). Another consequence is quite bizarre: objects with half-integer spin must be rotated by 720° (not 360°) before they return to their initial state! This peculiar behaviour has been demonstrated using very slow (ultracold) spin-oriented neutrons from a reactor, which are split into two beams. In one beam the neutrons are rotated about an axis along their direction of motion through some angle, and then the beams are recombined. It is found that when the rotation angle is 360°, the combined beams are out of phase and cancel, (meaning that they are shifted away from the detector) while after 720° of rotation the beams are in phase and reinforced (meaning that they show a large signal at the detector). A rotation of 720° is needed to put the neutrons back in their original state.

## Charge, Parity, and Time Reversal (CPT) Symmetry

Three other symmetry principles important in nuclear science are parity P, time reversal invariance T, and charge conjugation C. They deal with the questions, respectively, of whether a nucleus behaves in a different way if its spatial configuration is reversed (P), if the direction of time is made to run backwards instead of forward (T), or if the matter particles of the nucleus are changed to antimatter (C). All charged particles with spin 1/2 (electrons, quarks, etc.) have antimatter counterparts of opposite charge and of opposite parity. Particle and antiparticle, when they come together, can annihilate, disappearing and releasing their total mass energy in some other form, most often gamma rays.

The changes in symmetry properties can be thought of as "mirrors" in which some property of the nucleus (space, time, or charge) is reflected or reversed. A real mirror reflection provides a concrete example of this because mirror reflection reverses the space direction perpendicular to the plane of the mirror. As a consequence, the mirror image of a right-handed glove is a left-handed glove. This is in effect a parity transformation (although a true P transformation should reverse all three spatial axes instead of only one).

Until 1957 it was believed that the laws of physics were invariant under parity transformations and that no physics experiment could show a preference for left-handedness or right-handedness. Inversion, or mirror, symmetry was expected of nature. It came as some surprise that parity, P, symmetry is broken by the radioactive decay beta decay process. C. S. Wu and her collaborators found that when a specific nucleus was placed in a magnetic field, electrons from the beta decay were preferentially emitted in the direction opposite that of the aligned angular momentum of the nucleus. When it is possible to distinguish these two cases in a mirror, parity is not conserved. As a result, the world we live in is distinguishable from its mirror image.

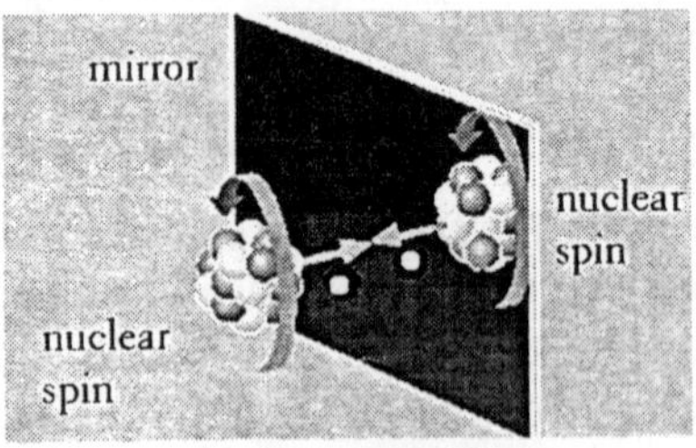


**Fig.** Symmetry Rotation of Nuclei

The direction of the emitted electron (arrow) reverses on mirror reflection, but the direction of rotation (angular momentum) is not changed. Thus the nucleus before the mirror represents the actual directional preference, while its mirror reflection represents a directional preference not found in nature. A physics experiment can therefore distinguish between the object and its mirror image.

If, however, we made a nucleus out of antimatter (antiprotons and antineutrons) its beta decay would behave in the same way, except that the mirror image would represent the preferred direction of electron emission, while the antinucleus in front of the mirror would represent a directional preference not found in nature.

The great physicist, Richard Feynman, told a story to illustrate this point: suppose you were in two-way contact with some alien species, but only by "telegraph" (*i.e.*, light flashes or radio signals). The well known procedures of SETI (Search for Extraterrestrial Intelligence), starting with prime numbers and progressing to pictures, physics, and chemistry information could be used to develop a common language and arrive at a good level of communication. You could tell the alien how tall you are by expressing your height in mutually understood wavelengths of light. You could tell the alien how old you are as some large number of ticks of a light-frequency clock. Now you want to explain how humans shake hands when they meet, and you describe extending your right hand. "Wait a moment!" says the alien. "What do you mean by 'right'?"

Feynman also had a punch line to this story. Suppose, after lots of communication you finally can go into space and meet your alien counterpart. If, as you approach one another, the alien extends its left hand to shake, watch out! He's made of antimatter! This, of course, is because a parity violation experiment constructed of antimatter would give the opposite result.

If the mirror not only reversed spatial direction but also changed matter to antimatter, then the experiment in front of the mirror would look just like its mirror image. Changing both C and P preserves the symmetry and we call this CP symmetry. The separate violations of P symmetry and C symmetry cancel to preserve CP symmetry. These symmetry violations arise only from the weak interaction, not from the strong and electromagnetic interactions, and therefore shows up strongly only in beta decay.

There are fundamental reasons for expecting that nature at a minimum has CPT symmetry–that no asymmetries will be found after reversing charge, space, and time. Therefore, CP symmetry implies T symmetry (or time-reversal invariance). One can demonstrate this symmetry by asking the following question. Suppose you had a movie of some physical process. If the movie were run backwards through the projector, could you tell from the images on the screen that the movie was running backwards? Clearly in everyday life there would be no problem in telling the difference. A movie of a street scene, an egg hitting the floor, or a dive into a swimming pool has an obvious "time arrow" pointing from the past to the future. But at the atomic level there are no obvious clues to time direction. An electron orbiting an atom or even making a quantum jump to produce a photon looks like a valid physical process in either time direction. The everyday "arrow of time" does not seem to have a counterpart in the microscopic world–a problem for which physics currently has no answer.

Until 1964 it was thought that the combination CP was a valid symmetry of the Universe. That year, Christenson, Cronin, Fitch and Turlay observed the decay of the long-lived neutral K meson, $K_L^e$, to $\pi^+ + \pi^-$. If CP were a good symmetry, the $K_L^e$ would have CP = -1 and could only decay to three pions, not two. Since the experiment observed the two pion decay, they showed that the symmetry CP could be violated. If CPT symmetry is to be preserved, the CP violation must be compensated by a violation of time reversal invariance. Indeed later experiments with $K_0$ systems showed direct T violations, in the sense that certain reaction processes involving K mesons have a different probability in the forward time direction ($A + B \rightarrow C + D$) from that in the reverse time direction ($C + D \rightarrow A + B$). Nuclear physicists have conducted many investigations searching for similar T violations in nuclear decays and reactions, but at this time none have been found.

This may change soon. Time reversal invariance implies that the neutron can have no electric dipole moment, a property implying separation of internal charges and an external electric field with its lines in loops like Earth's magnetic field. Currently ultracold neutrons are being used to make very sensitive tests of the neutron's electric dipole moment, and it is anticipated that a nonzero value may be found within the next few years.

## Matter and Antimatter

Time-reversal invariance and the CP violation are connected to another asymmetry of the universe, the imbalance between matter and antimatter. At the microscopic level, matter and antimatter are always created together in 1:1 correspondence. High energy collisions produce equal numbers of quarks and antiquarks. And yet, our universe has a conspicuous surplus of matter, of which we and our surroundings are made. How did this happen?

A clue to this deep mystery is provided by the CP violation in the $K^0$ meson, which shows decay modes having a preference for matter over antimatter. The $K^0$ does not have enough mass for its decay to produce protons, but its decay asymmetry suggests that some more massive particle, perhaps a $B^0$ meson containing a bottom quark, might in the early universe have decayed preferentially into protons rather than antiprotons, leading to the present day dominance of matter. Future experiments using the B-Factory, presently under construction at the Stanford Linear Accelerator Centre (SLAC), will investigate this problem.

Antimatter exists in nature only in the form of antiprotons present in very small numbers in cosmic rays and in positrons (antimatter electrons) produced in some radioactive decays. Recently, evidence has also been found for a "fountain" of positrons ejected from some object near the centre of our galaxy, presumably a black hole.

However, we are getting better and better at producing and storing antimatter in the laboratory. Antiprotons, antineutrons, and even antideuterons (a nucleus consisting of an antineutron and an antiproton) are routinely produced using high energy particle accelerators at Fermilab in Illinois and CERN in Geneva, Switzerland. Positrons and antiprotons have been trapped in electric and magnetic fields and held under high vacuum for several months. Recently, "antihydrogen" atoms having a positron orbiting an antiproton have been formed in laboratory experiments. These researchers are looking for any indication that trapped positrons, antiprotons, and antihydrogen atoms show a behaviour that differs in any way from that of their normal matter counterparts, because any such difference would represent a violation of CPT symmetry.

Antimatter nuclei are also interesting for other reasons. Special facilities at CERN and Fermilab provide beams of low energy antiprotons and permit nuclear scientists to study the interactions of antiprotons with matter. While a positron and an electron usually annihilate to form a pair of gamma-ray photons traveling in opposite directions, the annihilation of an antiproton with a proton is more complicated. Several $\pi$ mesons are usually produced. About a third of the mass energy of the proton-antiproton pair becomes inaccessible in the form of energetic neutrinos.

Nevertheless, antimatter can be viewed as an extremely compact form of stored energy that can be released at will by annihilation with matter. The US Air Force has commissioned design studies of antimatter-powered space vehicles that, given a supply of antimatter, look quite feasible. The problem with such schemes is that production of any significant quantity of antimatter would cost far too much right now to be economically feasible.

## Other Symmetries

In addition to the symmetries described above, nuclear scientists use a number of other approximate symmetries to describe and predict the behaviour of nuclei. Examples of these are charge independence, the expectation that, at the nuclear level, neutron-proton systems should behave the same as proton-proton or neutron-neutron systems; and charge symmetry, the expectation that the interactions between two neutrons should be the same as that between two protons. Charge symmetry can be demonstrated by comparing "mirror nuclei," two low-mass nuclei that have their neutron and proton numbers interchanged and which have very similar nuclear structure$^{13}$C (Z=6, N=7) and $^{13}$N (Z=7, N=6).

A related symmetry is isospin symmetry, which is related to interchanging the roles of neutrons and protons in certain nuclei. These three symmetries are destroyed when the Coulomb force becomes sufficiently strong, but nevertheless they have proved to be useful approximations in many areas of nuclear science.

# Chapter 7

# Heavy Elements

When a nucleus captures a neutron, it often tries to correct for its neutron excess by beta decay, turning a neutron into a proton and thus creating an atom with atomic number Z increased by one unit. This commonly observed phenomenon suggests a way to create new elements of increased atomic number and thus to create ever more massive elements that are not found on Earth. Most of these elements are radioactive, with very short half-lives. However, theories of nuclear structure predict that at a certain atomic number, which is currently beyond present experimental limits, new long-lived nuclei can be created.

The most massive naturally occurring element on Earth is uranium (U), with a nucleus of 92 protons. In 1934, scientists started the search for more massive elements with 93 or more protons. They succeeded in 1940 when neptunium (Np, $Z = 93$) was synthesized at the University of California, Berkeley. Edwin McMillan and Philip Abelson observed Np while studying fission products produced in the bombardment of $^{238}U$ with thermal neutrons. They found a radioactive reaction product that was not a fission product. This product was formed by the capture of a neutron to produce $^{239}U$, which subsequently $\beta^-$ decayed to $^{239}Np$ via the reaction

$$^{239}U + {}^{1}n \rightarrow {}^{239}U \text{ (23.5 min)} + \beta \rightarrow {}^{239}Np \text{ (2.36 d)},$$

where the half life of the nucleus is indicated in the parenthesis. McMillan and Abelson chemically separated this new element, Np, from the interfering fission products and chemically identified it as neptunium. Since this breakthrough discovery, scientists from all over the world have been trying to discover ever more massive artificially produced elements.

Plutonium (Pu, Z = 94) was discovered in 1941 by bombarding a uranium target with deuterons (a hydrogen nucleus with one proton and one neutron) in the 60-Inch Cyclotron at Berkeley. Glenn Seaborg, Arthur Wahl and Joseph Kennedy chemically separated neptunium from the target and detected alpha particles from the plutonium daughter nuclei, as:

$$^{238}U + {}^{2}H + 2\ {}^{1}n + {}^{238}Np\ (2.1\ d) \rightarrow {}^{238}Pu\ (88\ yr).$$

They chemically identified the isotope $^{238}Pu$. Then, joined by Emilio Segre, they identified $^{239}Pu$ and showed that it was fissionable with thermal neutrons.

Once $^{239}Pu$ was discovered, there was the potential for using it as a new target to produce more massive elements. In 1944 a $^{239}Pu$ target was bombarded with alpha particles at the 60-Inch Cyclotron to produce curium (Cm, Z = 96) through the following nuclear reaction

$$^{239}Pu\ (24{,}110\ yr) + {}^{4}He \rightarrow {}^{242}Cm\ (162.8\ d) + {}^{1}n.$$

After bombardment the material was sent to the Metallurgical Laboratory at The University of Chicago for chemical separation and identification of the new element. $^{242}Cm$ decays to $^{238}Pu$ by emitting alpha particles. The identification of curium was possible because the alpha decay of the daughter nucleus, $^{238}Pu$, was already known and could be used as a signature for the identification of the curium precursor.

The discovery of americium (Am, Z = 95) soon followed when a $^{239}Pu$ target was bombarded with thermal neutrons in a nuclear reactor. Plutonium captured several neutrons and ultimately became americium ($^{241}Am$):

$^{239}Pu$ (24,110 yr) + $^{1}n$ → $^{240}Pu$ (6,564 yr) + $^{1}n$ → $^{241}Pu$ (14.4 yr), and $^{241}Pu$ (14.4 yr) $\rightarrow$ $^{241}Am$ (432.7 yr) + $^{1}n$ $\rightarrow$ $^{242}Am$ (16.0 h) $\rightarrow$ $^{242}Cm$ (162.8 d).

Americium was chemically separated from plutonium and further identified by observing its beta decay to the known $^{242}Cm$ isotope.

Once americium and curium were discovered and isolated in macroscopic amounts, they were used as targets to produce more massive elements through particle bombardments. Berkelium (Bk, Z = 97) was produced by bombarding milligram quantities of americium with helium ions,

$$^{241}Am \text{ (432.7 yr)} + {}^{4}He \rightarrow {}^{243}Bk \text{ (4.5 h)} + 2\,{}^{1}n.$$

Rapid chemical techniques were developed in order to separate and identify this new short-lived element. Likewise, californium (Cf, Z = 98) was produced in a helium bombardment of a target made of microgram amounts of curium by

$$^{242}Cm \text{ (162.8 d)} + {}^{4}He \rightarrow {}^{245}Cf \text{ (45 min)} + {}^{1}n.$$

The identification of this element was accomplished with only the 5000 atoms produced in this experiment. The next two elements, einsteinium (Es, Z = 99) and fermium (Fm, Z = 100), were unexpectedly found in the debris from the "Mike" thermonuclear explosion that took place in the Pacific Ocean in 1952. Debris from the explosion was collected and analysed at several laboratories, and the new elements were discovered in chemical separations of the material. Scientists explained the production of einsteinium and fermium through multiple neutron captures by the uranium used in the thermonuclear device followed by several successive beta decays, which ultimately resulted in atoms with atomic numbers 99 and 100.

The last three elements in the actinide series are mendelevium (Md, Z = 101), nobelium (No, Z = 102) and lawrencium (Lr, Z = 103). Mendelevium was truly a unique discovery because the new element was produced and identified virtually one atom at a time. Einsteinium was bombarded with helium ions to produce mendelevium through:

$$^{253}Es\ (20.5\ d) + {}^{4}He \rightarrow {}^{256}Md\ (78.1\ min) + {}^{1}n.$$

The production of mendelevium was estimated to be only a few atoms per experiment. The reaction products from the bombardment were collected on thin gold foils that were dissolved in an acid solution, and then chemically treated in order to separate and identify the Md atoms. This is commonly called the recoil method, and is used when small amounts of atoms are produced.

The discovery of nobelium was controversial. A team of scientists from several different laboratories claimed discovery in 1957. However, scientists from the United States and the Soviet Union could not confirm their findings. The original claim was proven to be false; the product that was thought to be nobelium was actually something completely different. Nobelium was finally produced and positively identified in 1958 through the following reaction:

$$^{246}Cm\ (4730\ yr) + {}^{12}C \rightarrow {}^{254}No\ (55\ s) + 4\ {}^{1}n.$$

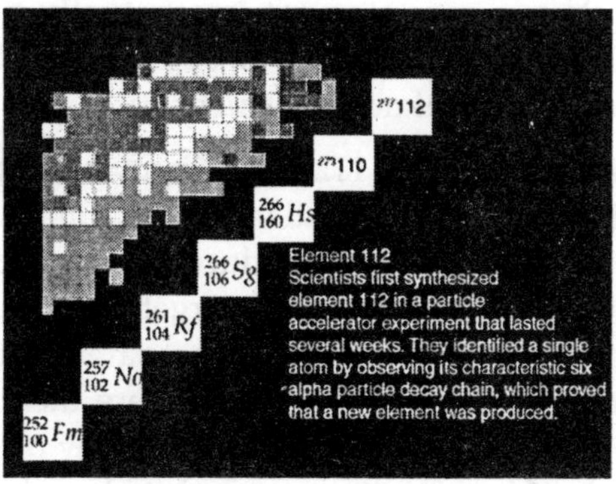


**Fig.** Heavy Element Linear Accelerator

The first identification of lawrencium was made at the Berkeley Lab's Heavy Ion Linear Accelerator (HILAC) in 1961. Several targets of californium isotopes were bombarded with beams of boron. The reaction products were collected on a mylar tape and moved past a series of alpha detectors. The element lawrencium was identified on the basis of the known alpha decays of its descendant nuclei.

Once the actinide series was filled in the periodic table, work was started to produce the transactinide elements. The most recent version of the periodic table includes transactinide elements from rutherfordium (Rf, Z = 104) to element 112, which has not yet been named. As scientists tried to discover and identify the transactinide elements, they encountered several problems that made their task extremely difficult.

The rates for producing these elements in accelerator bombardments were incredibly low. Consequently, only a few atoms could be produced over the course of the entire experiment. Sometimes these experiments went on continuously for several weeks. In addition, the half-lives of the transactinides are very short, making their identification a difficult challenge. Unlike neptunium and plutonium, these elements may undergo radioactive decay before being detected. As a result, chemical separations are generally very difficult to perform for identifying transactinides. In order to confirm that a new element has actually been made, members of its decay chain must be detected and traced back as a signature to the unknown element.

Elements 104 through 112 were discovered by observing the decay chain of their descendant nuclei. Only a few atoms of these new elements were produced in each experiment. The atoms were isolated from the target and beam material by using a particle separator which separates atoms based on their different masses. The atoms were then allowed to decay and the subsequent alpha particle decay products from the descendant nuclei were correlated to identify the unknown parent nucleus. In the discovery of element 112, the individual atoms were identified by the decay chains of parent-daughter-granddaughter nuclei. Different combinations of target and projectile were used in accelerators to produce these new elements.

Rutherfordium (104), dubnium (105), and seaborgium (106) were synthesized and identified at Berkeley. Bohrium (107), hassium (108), and meitnerium (109) were synthesized and identified in the early 1980s at the Gesellschaft für

Schwerionenforschung (GSI) laboratory near Darmstadt, Germany. In the 1990s, unnamed element 110 was identified at the GSI, Berkeley, and Dubna laboratories, and elements 111 and 112 were found at the GSI laboratory. The names for these new elements are under discussion.

New experimental techniques and apparatus are being developed for scientists to extend the periodic table to even more massive elements. A more efficient particle separator, will use magnetic fields to separate atoms based on their mass and charge. This equipment will allow detection of nuclides with low production rates and extremely short half-lives. There are plans for using such a separator to produce elements 114 and 116. The present limits for discovering new elements are based on the low production rates and short half-lives. The hope is that new development in detection equipment will increase the sensitivity for detecting fewer atoms (or even a single atom) with very short half-lives. A widely used theory predicts the existence of elements up to Z = 125. In this model, more massive elements would be so unstable that their nuclei would immediately fly apart into nucleons and nuclear fragments and never become atoms.

Traditionally, the discoverers of a new element chose its name. Then the International Union of Pure and Applied Chemistry (IUPAC) officially approves the proposed name. However, for several of the transfermium elements (Z = 102 and higher) there had been competing claims to their discovery. Recently, the names for elements have been decided up to Z = 109.

It has been possible to study the chemical properties on the macroscopic scale for elements as massive as einsteinium (element 99) and on the tracer scale for elements as massive as seaborgium (element 106). The elements beyond the actinides in the Periodic Table are termed the "transactinides" and are shown in a Modern Periodic Table with all of the undiscovered elements through number 118 in their expected places.

**Table: "Heavy" Element Names that are Approved by IUPAC.**

| Element Number | IUPAC Proposal | Symbol |
|---|---|---|
| 101 | Mendelevium | Md |
| 102 | Nobelium | No |
| 103 | Lawrencium | Lr |
| 104 | Rutherfordium | Rf |
| 105 | Dubnium | Db |
| 106 | Seaborgium | Sg |
| 107 | Bohrium | Bh |
| 108 | Hassium | Hs |
| 109 | Meitnerium | Mt |

The yields of the most massive elements produced in bombardments of target nuclei with "heavy" ions become extremely small with increasing atomic number, dropping to as little as one atom per week of bombardment for elements as massive as atomic number 112. The half-lives decrease into the millisecond and the microsecond range so that identification of the new nuclei becomes increasingly difficult. These half-lives would be impossibly short were it not for the presence of closed shells of nucleons to increase nuclear stability.

It has been possible to study the chemical properties of rutherfordium, hahnium and seaborgium using the advanced techniques of one-atom-at-a-time chemistry, with the result that these properties in the main are consistent with those expected on the basis of extrapolation from those of their lower-mass *homologues* in the Periodic Table, hafnium, tantalum, and tungsten. It should be possible to similarly study the chemical properties of bohrium (107) and hassium (108), the homologues of rhenium and osmium. However, the chemical properties cannot be determined reliably in detail from trends exhibited by the lighter homologues, because of the important role played by relativistic effects (the fact that some electrons are moving at velocities near the speed of light) in these more massive elements.

It will surely be possible to study the macroscopic properties of fermium (element 100) and not out of the question that this will eventually be done for mendelevium (element 101). The art of one-atom-at-a-time chemistry will advance far beyond what can be imagined today to make it possible to study the chemistry of most massive elements. All of this will result in the delineation of relativistic effects on the chemical properties of these very massive elements, which might thus be substantially different from those expected by simple extrapolation from their less massive homologues in the Periodic Table.

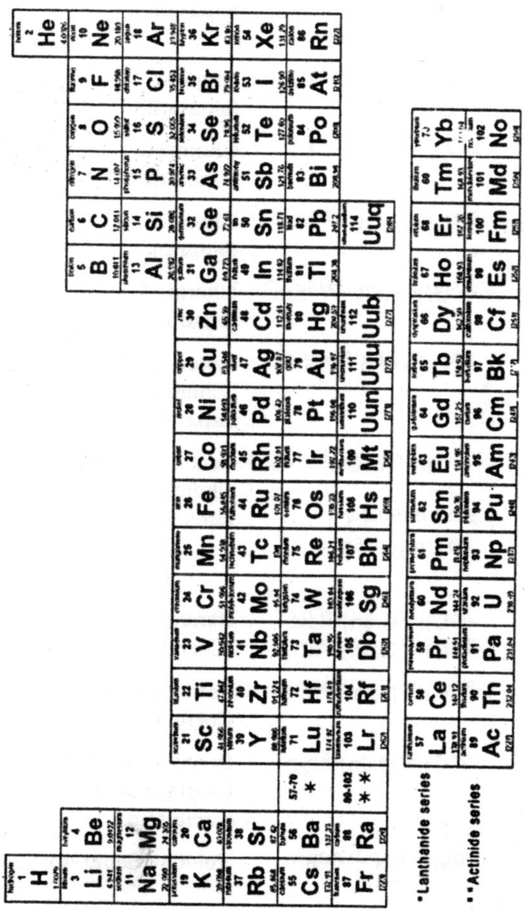


Fig. Periodic Table

**Table: Summary of The Reactions and Methods used in The Discovery of The Actinide and Transactinide Elements.**

| Element | Production Reaction(s)[1] | Method of Discovery | Year |
|---|---|---|---|
| neptunium (Np) | $^{238}U(n,\beta^-)^{239}Np$ | Chemical Separation | 1940 |
| plutonium (Pu) | $^{238}U(^{2}H,2n)^{238}Np$<br>$^{238}Np(\beta^-)^{238}Pu$ | Chemical Separation | 1941 |
| curium (Cm) | $^{239}Pu(^{4}He,n)^{242}Cm$ | Chemical Separation | 1944 |
| americium (Am) | $^{239}Pu(n,\gamma)^{240}Pu$<br>$^{240}Pu(n,\gamma)^{241}Pu$<br>$^{241}Pu(\beta^-)^{241}Am$ | Chemical Separation | 1945 |
| berkelium (Bk) | $^{241}Am(^{4}He,2n)^{243}Bk$ | Chemical Separation | 1949 |
| californium (Cf) | $^{242}Cm(^{4}He,n)^{245}Cf$ | Chemical Separation | 1950 |
| einsteinium (Es) | $^{238}U(15n,7\beta^-)^{253}Es$[2] | Chemical Separation | 1952 |
| fermium (Fm) | $^{238}U(17n,8\beta^-)^{255}Fm$[2] | Chemical Separation | 1953 |
| mendelevium (Md) | $^{253}Es(^{4}He,n)^{256}Md$ | Recoil Method, Chemical Separation | 1955 |
| nobelium (No) | $^{246}Cm(^{12}C,4n)^{254}No$ | Recoil Method, Chemical | 1958 |
| lawrencium (Lr) | $^{249/250/251/252}Cf(^{10/11}B,xn)^{258}Lr$[3] | Direct α Counting | 1961 |
| rutherfordium (Rf) | $^{249}Cf(^{12}C,4n)^{257}Rf$<br>$^{249}Cf(^{13}C,3n)^{259}Rf$ | Parent-Daughter α Correlation | 1969 |
| dubnium (Db) | $^{249}Cf(^{15}N,4n)^{260}Db$ | Parent-Daughte α Correlation | 1970 |
| seaborgium (Sg) | $^{249}Cf(^{18}O,4n)^{263}Sg$ | Parent-Daughter-Granddaughter α Correlation | 1974 |
| bohrium (Bh) | $^{209}Bi(^{54}Cr,n)^{262}Bh$ | Velocity Separator[4] | 1981 |

| | | | |
|---|---|---|---|
| hassium (Hs) | $^{208}Pb(^{58}Fe,n)^{265}Hs$ | Velocity Separator[4] | 1984 |
| meitnerium (Mt) | $^{209}Bi(^{58}Fe,n)^{266}Mt$ | Velocity Separator[4] | 1982 |
| 110 (unnamed) | $^{209}Bi(^{59}Co,n)^{267}110$[5] | Mass Separator | 1991 |
| | $^{208}Pb(^{62,64}Ni,n)^{269,271}110$[5,6] | Velocity Separator[4] | 1994 |
| | $^{244}Pu(^{34}S,5n)^{273}110$[5] | Recoil Separator | 1995 |
| 111 (unnamed) | $^{209}Bi(^{64}Ni,n)^{272}111$ | Velocity Separator[4] | 1994 |
| 112 (unnamed) | $^{208}Pb(^{70}Zn,n)^{277}112$ | Velocity Separator[4] | 1996 |

1. When more than one reaction is given, it means that a sequence of reactions was necessary to discover the element.
2. The Es and Fm reactions were not done in a laboratory setting; multiple neutron captures, followed by successive beta decays were the result of a thermonuclear explosion.
3. A mixture of four Cf isotopes were bombarded simultaneously with a beam containing 10B and 11B. The xn means that different numbers of neutrons were emitted, depending on the actual combination of target and beam used to produce Lr.
4. A velocity separator is used to separate reaction products based on the fact that reaction products of different masses will be emitted with different velocities.
5. Three different claims to element 110 have been made by teams from Lawrence Berkeley National Laboratory, the GSI Laboratory in Germany and a collaboration from Dubna in Russia and Lawrence Livermore National Laboratory. As of now, no one claim has been given sole credit for the discovery.
6. At GSI two different reactions were used to produce two different isotopes of element 110.

## Chapter 8

# Radioactivity

### STABILITY OF NUCLEI

As mass numbers become larger, the ratio of neutrons to protons in the nucleus becomes larger for the stable nuclei. Non-stable nuclei may have an excess or deficiency of neutrons and undergo a transformation process known as beta (β) decay. Non-stable nuclei can also undergo a variety of other processes such as alpha (α) or neutron (n) decay. As a result of these decay processes, the final nucleus is in a more stable or more tightly bound configuration.

In radioactive processes, particles or electromagnetic radiation are emitted from the nucleus. The most common forms of radiation emitted have been traditionally classified as alpha (α), beta (β), and gamma (γ) radiation. Nuclear radiation occurs in other forms, including the emission of protons or neutrons or spontaneous fission of a massive nucleus.

Of the nuclei found on Earth, the vast majority are stable. This is so because almost all short-lived radioactive nuclei have decayed during the history of the Earth. There are approximately 270 stable isotopes and 50 naturally occurring radioisotopes (radioactive isotopes). Thousands of other radioisotopes have been made in the laboratory.

Radioactive decay will change one nucleus to another if the product nucleus has a greater nuclear binding energy than the initial decaying nucleus. The difference in binding energy (comparing the before and after states) determines which decays are energetically possible and which are not. The excess binding energy appears as kinetic energy or rest mass energy of the decay products.

The Chart of the Nuclides, part of which is shown above is a plot of nuclei as a function of proton number, *Z*, and neutron number, *N*. All stable nuclei and known radioactive nuclei, both naturally occurring and manmade, are shown on this chart, along with their decay properties. Nuclei with an excess of protons or neutrons in comparison with the stable nuclei will decay toward the stable nuclei by changing protons into neutrons or neutrons into protons, or else by shedding neutrons or protons either singly or in combination. Nuclei are also unstable if they are excited, that is, not in their lowest energy states. In this case the nucleus can decay by getting rid of its excess energy without changing *Z* or *N* by emitting a gamma ray.

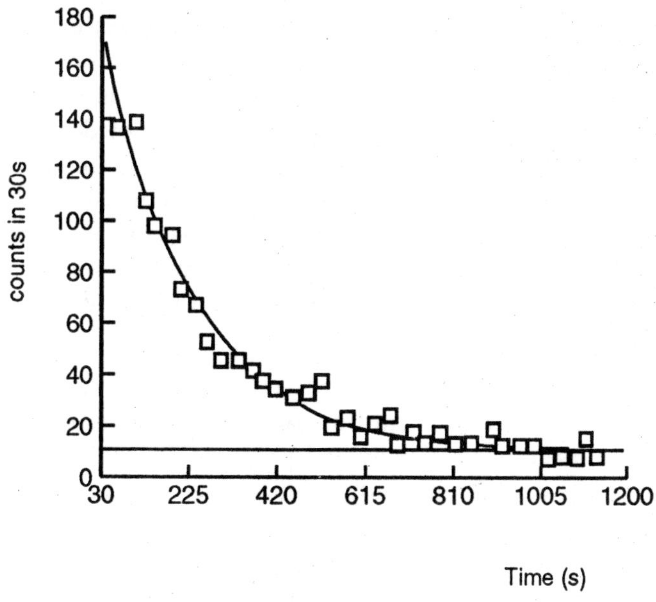


Nuclear decay processes must satisfy several conservation laws, meaning that the value of the conserved quantity after the decay, taking into account all the decay products, must equal the same quantity evaluated for the nucleus before the decay. Conserved quantities include total energy (including mass), electric charge, linear and angular momentum, number of nucleons, and lepton number (sum of the number of electrons, neutrinos, positrons and antineutrinos–with antiparticles counting as –1).

137Ba decay data, counting numbers of decays observed in 30-second intervals. The best-fit exponential curve is shown. The points do not fall exactly because of statistical counting fluctuations.

The probability that a particular nucleus will undergo radioactive decay during a fixed length of time does not depend on the age of the nucleus or how it was created. Although the exact lifetime of one particular nucleus cannot be predicted, the mean (or average) lifetime of a sample containing many nuclei of the same isotope can be predicted and measured. A convenient way of determining the lifetime of an isotope is to measure how long it takes for one-half of the nuclei in a sample to decay–this quantity is called the half-life, $t_{1/2}$. Of the original nuclei that did not decay, half will decay if we wait another half-life, leaving one-quarter of the original sample after a total time of two half-lives. After three half-lives, one-eighth of the original sample will remain and so on. Measured half-lives vary from tiny fractions of seconds to billions of years, depending on the isotope.

The number of nuclei in a sample that will decay in a given interval of time is proportional to the number of nuclei in the sample. This condition leads to radioactive decay showing itself as an exponential process. The number, $N$, of the original nuclei remaining after a time $t$ from an original sample of $N_0$ nuclei is

$$N = N_0 e^{-(t/T)}$$

where $T$ is the mean lifetime of the parent nuclei. From this relation, it can be shown that $t_{1/2} = 0.693T$.

## Nuclear Decay

Whenever a nucleus can attain a more stable (i.e., more tightly bound) configuration by emitting radiation, a spontaneous disintegration process known as radioactive decay or nuclear decay may occur. In practice, this "radiation" may be electromagnetic radiation, particles, or both. Detailed studies of radioactive decay and nuclear reaction processes have led to the formulation of useful conservation principles. The four principles of most interest in this module are:

*Conservation of electric charge* implies that charges are neither created nor destroyed. Single positive and negative charges may, however, neutralize each other. It is also possible for a neutral particle to produce one charge of each sign.

*Conservation of mass number* does not allow a net change in the number of nucleons. However, the conversion of a proton to a neutron and vice versa is allowed.

*Conservation of mass and energy* implies that the total of the kinetic energy and the energy equivalent of the mass in a system must be conserved in all decays and reactions. Mass can be converted to energy and energy can be converted to mass, but the sum of mass and energy must be constant.

*Conservation of momentum* is responsible for the distribution of the available kinetic energy among product nuclei, particles, and/or radiation. The total amount is the same before and after the reaction even though it may be distributed differently among entirely different nuclides and/or particles.

## Alpha Decay

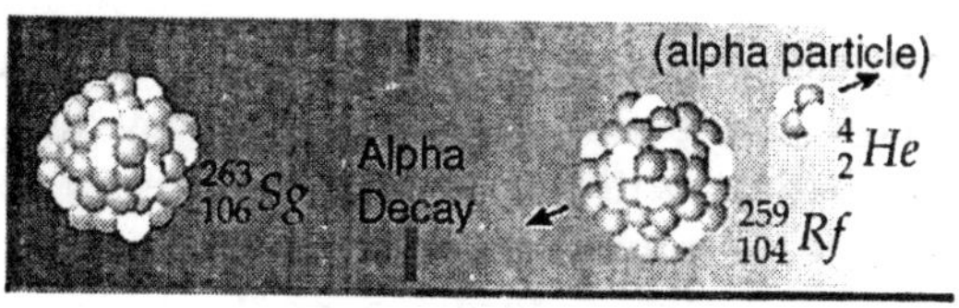


Fig. An alpha-particle decay

In alpha decay, the nucleus emits a 4He nucleus, an *alpha particle*. Alpha decay occurs most often in massive nuclei that have too large a proton to neutron ratio. An alpha particle, with its two protons and two neutrons, is a very stable configuration of particles. Alpha radiation reduces the ratio of protons to neutrons in the parent nucleus, bringing it to a more stable configuration. Many nuclei more massive than lead decay by this method.

Consider the example of $^{210}$Po decaying by the emission of an alpha particle. The reaction can be written $_{210}$Po Æ $_{206}$Pb + $_{4}$He. This polonium nucleus has 84 protons and 126 neutrons. The ratio of protons to neutrons is $Z/N$ = 84/126, or 0.667. A $_{206}$Pb nucleus has 82 protons and 124 neutrons, which gives a ratio of 82/124, or 0.661. This small change in the Z/N ratio is enough to put the nucleus into a more stable state, brings the "daughter" nucleus (decay product) into the region of stable nuclei in the Chart of the Nuclides.

In alpha decay, the atomic number changes, so the original (or parent) atoms and the decay-product (or daughter) atoms are different elements and therefore have different chemical properties.

In the alpha decay of a nucleus, the change in binding energy appears as the kinetic energy of the alpha particle and the daughter nucleus. Because this energy must be shared between these two particles, and because the alpha particle and daughter nucleus must have equal and opposite momenta, the emitted alpha particle and recoiling nucleus will each have a well-defined energy after the decay. Because of its smaller mass, most of the kinetic energy goes to the alpha particle.

## Beta Decay

Beta particles are electrons or positrons (electrons with positive electric charge, or antielectrons). Beta decay occurs when, in a nucleus with too many protons or too many neutrons, one of the protons or neutrons is transformed into the other. In beta minus decay, a neutron decays into a proton, an electron, and an antineutrino: n → p + e – +. In beta plus decay, a proton decays into a neutron, a positron, and a neutrino: p → n + e+ +*n*. Both reactions occur because in different regions of the Chart of the Nuclides, one or the other will move the product closer to the region of stability.

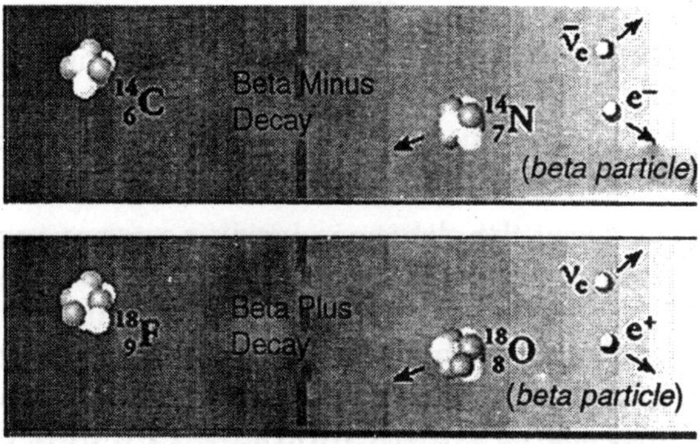


**Fig.** Beta Decays

These particular reactions take place because conservation laws are obeyed. Electric charge conservation requires that if an electrically neutral neutron becomes a positively charged proton, an electrically negative particle (in this case, an electron) must also be produced. Similarly, conservation of lepton number requires that if a neutron (lepton number = 0) decays into a proton (lepton number = 0) and an electron (lepton number = 1), a particle with a lepton number of -1 (in this case an antineutrino) must also be produced. The leptons emitted in beta decay did not exist in the nucleus before the decay–they are created at the instant of the decay.

To the best of our knowledge, an isolated proton, a hydrogen nucleus with or without an electron, does not decay. However within a nucleus, the beta decay process can change a proton to a neutron. An isolated neutron is unstable and will decay with a half-life of 10.5 minutes. A neutron in a nucleus will decay if a more stable nucleus results; the half-life of the decay depends on the isotope. If it leads to a more stable nucleus, a proton in a nucleus may capture an electron from the atom (electron capture), and change into a neutron and a neutrino.

Proton decay, neutron decay, and electron capture are three ways in which protons can be changed into neutrons or vice-versa; in each decay there is a change in the atomic number, so that the parent and daughter atoms are different elements. In all three processes, the number *A* of nucleons remains the same, while both proton number, *Z*, and neutron number, *N*, increase or decrease by 1.

In beta decay the change in binding energy appears as the mass energy and kinetic energy of the beta particle, the energy of the neutrino, and the kinetic energy of the recoiling daughter nucleus. The energy of an emitted beta particle from a particular decay can take on a range of values because the energy can be shared in many ways among the three particles while still obeying energy and momentum conservation.

## Gamma Decay

**Fig.** Gamma Decay

In gamma decay, a nucleus changes from a higher energy state to a lower energy state through the emission of electromagnetic radiation (photons). The number of protons (and neutrons) in the nucleus does not change in this process, so the parent and daughter atoms are the same chemical element. In the gamma decay of a nucleus, the emitted photon and recoiling nucleus each have a well-defined energy after the decay. The characteristic energy is divided between only two particles.

## The Discovery of Radioactivity

In 1896 Henri Becquerel was using naturally fluorescent minerals to study the properties of x-rays, which had been discovered in 1895 by Wilhelm Roentgen. He exposed potassium uranyl sulfate to sunlight and then placed it on photographic plates wrapped in black paper, believing that the uranium absorbed the sun's energy and then emitted it as x-rays. This hypothesis was disproved on the $26^{th}$–$27^{th}$ of February, when his experiment "failed" because it was overcast in Paris. For some reason, Becquerel decided to develop his photographic plates anyway. To his surprise, the images were strong and clear, proving that the uranium emitted radiation without an external source of energy such as the sun. Becquerel had discovered radioactivity.

Becquerel used an apparatus similar to that displayed below to show that the radiation he discovered could not be x-rays. X-rays are neutral and cannot be bent in a magnetic field. The new radiation was bent by the magnetic field so that the radiation must be charged and different than x-rays. When different radioactive substances were put in the magnetic field, they deflected in different directions or not at all, showing that there were three classes of radioactivity: negative, positive, and electrically neutral.

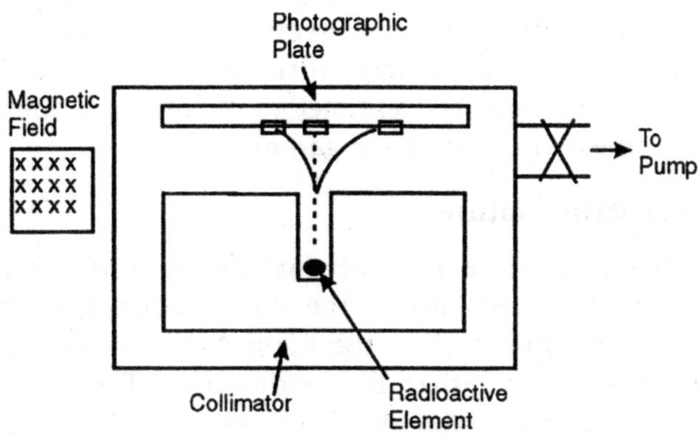


**Fig.** Radio Activity

The term radioactivity was actually coined by Marie Curie, who together with her husband Pierre, began investigating the phenomenon recently discovered by Becquerel. The Curies extracted uranium from ore and to their surprise, found that the leftover ore showed more activity than the pure uranium. They concluded that the ore contained other radioactive elements. This led to the discoveries of the elements polonium and radium. It took four more years of processing tons of ore to isolate enough of each element to determine their chemical properties.

Ernest Rutherford, who did many experiments studying the properties of radioactive decay, named these alpha, beta, and gamma particles, and classified them by their ability to penetrate matter. When the air from the chamber was removed, the alpha source made a spot on the photographic plate. When air was added, the spot disappeared. Thus, only a few centimeters of air were enough to stop the alpha radiation.

Because alpha particles carry more electric charge, are more massive, and move slowly compared to beta and gamma particles, they interact much more easily with matter. Beta particles are much less massive and move faster, but are still electrically charged. A sheet of aluminum one millimeter thick or several meters of air will stop these electrons and positrons. Because gamma rays carry no electric charge, they can penetrate large distances through materials before interacting–several centimeters of lead or a meter of concrete is needed to stop most gamma rays.

**Radioactivity in Nature**

Radioactivity is a natural part of our environment. Present-day Earth contains all the stable chemical elements from the lowest mass (H) to the highest (Pb and Bi). Every element with higher Z than Bi is radioactive. The earth also contains several primordial long-lived radioisotopes that have survived to the present in significant amounts. $_{40}K$, with its 1.3 billion year half-life, has the lowest mass of these isotopes and beta decays to both $_{40}Ar$ and $_{40}Ca$.

Many isotopes can decay by more than one method. When actinium-226 (Z=89) decays, 83percent of the rate is through β --decay, $_{226}Ac \rightarrow {}_{226}Th$ + e- +, 17percent is through electron capture, 226Ac + e- $\rightarrow {}_{226}Fr + n$, and the remainder, 0.006percent, is through α-decay, $_{226}Ac \rightarrow {}_{222}Fr + {}_{4}He$.

Therefore from 100,000 atoms of actinium, one would measure on average 83,000 beta particles and 6 alpha particles (plus 100,000 neutrinos or antineutrinos). These proportions are known as branching ratios. The branching ratios are different for the different radioactive nuclei.

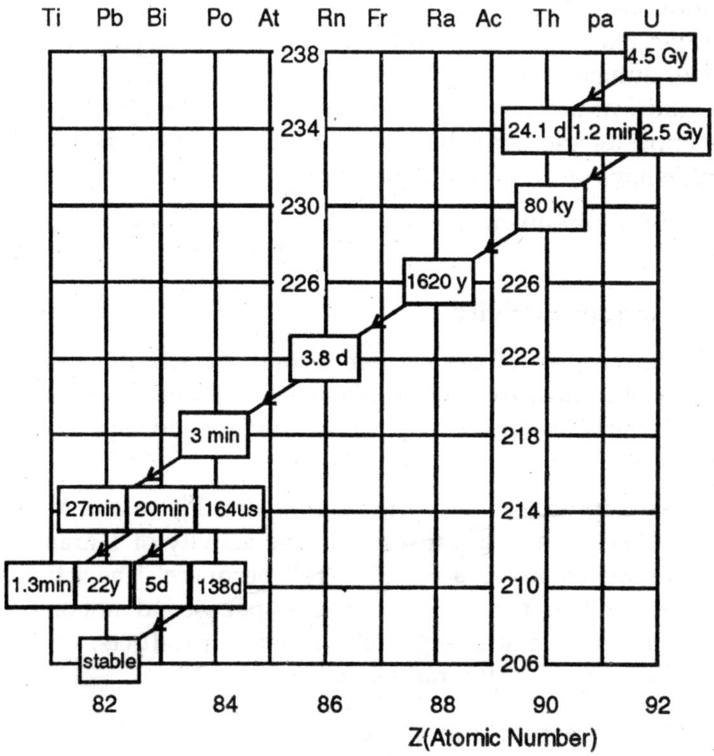


Three very massive elements, $_{232}$Th (14.1 billion year half-life), $_{235}$U (700 million year half-life), and $_{238}$U (4.5 billion year half-life) decay through complex "chains" of alpha and beta decays ending at the stable $_{208}$Pb, $_{207}$Pb, and $_{206}$Pb respectively. The decay chain for $_{238}$U. The ratio of uranium to lead present on Earth today gives us an estimate of its age (4.5 billion years). Given Earth's age, any much shorter lived radioactive nuclei present at its birth have already decayed into stable elements. One of the intermediate products of the $^{238}$U decay chain, $^{222}$Rn (radon) with a half-life of 3.8 days, is responsible for higher levels of background radiation in many parts of the world. This is primarily because it is a gas and can easily seep out of the earth into unfinished basements and then into the house.

Radioactive isotopes, $_{14}C$ and $_{7}Be$, are produced continuously through reactions of cosmic rays (high energy charged particles from outside Earth) with molecules in the upper atmosphere. $_{14}C$ is useful for radioactive dating. Also, the study of radioactivity is very important to understand the structure of the earth because radioactive decay heats the earth's interior to very high temperatures.

The uranium decay series. The vertical axis is atomic mass.

**Units of Radioactivity**

The number of decays per second, or activity, from a sample of radioactive nuclei is measured in becquerel (Bq), after Henri Becquerel. One decay per second equals one becquerel.

An older unit is the curie, named after Pierre and Marie Curie. One curie is approximately the activity of 1 gram of radium and equals (exactly) $3.7 \times 10^{10}$ becquerel. The activity depends only on the number of decays per second, not on the type of decay, the energy of the decay products, or the biological effects of the radiation.

**Four Fundamental Interactions**

The forces of gravity and electromagnetism are familiar in everyday life. Two new forces are introduced when discussing nuclear phenomena: the strong and weak interactions. When two protons encounter each other, they experience all four of the fundamental forces of nature simultaneously. The weak force governs beta decay and neutrino interactions with nuclei. The strong force, which we generally call the nuclear force, is actually the force that binds quarks together to form baryons (3 quarks) and mesons (a quark and an anti-quark). The nucleons of everyday matter, neutrons and protons, consist of the quark combinations *uud* and *udd*, respectively. The symbol *u* represents a single up quark, while the symbol *d* represents a single down quark.

The force that holds nucleons together to form an atomic nucleus can be thought to be a residual interaction between quarks inside each individual nucleon. This is analogous to what happens in a molecule. The electrons in an atom are bound to its nucleus by electromagnetism: when two atoms are relatively near, there is a residual interaction between the electron clouds that can form a covalent bond. The nucleus can thus be thought of as a "strong force molecule."

The force between two objects can be described as the exchange of a particle. The exchange particle transfers momentum and energy between the two objects, and is said to mediate the interaction. A simple analogue of this is a ball being thrown back and forth between two people. The momentum imparted to the ball by one person gets transferred to the other person when she catches the ball.

Strength and range of the four fundamental forces between two protons. Note that the strong force acts between quarks by an exchange of gluons. The residual strong force between two protons can be described by the exchange of a neutral pion. Note, the W± is not included as an exchange particle for the weak interaction because it is not exchanged in the simplest proton-proton interaction.

| **Interaction** | **Gravity** | **Weak** | **Electro-magnetism** | **Strong Residual** |
|---|---|---|---|---|
| Exchange particle | Graviton | Z0 | Photon | Pion |
| Mass $mc^2$ *(eV)* | 0 | $91\text{x}10^9$ | 0 | $135\text{x}10^6$ |
| Coupling constant $C^2$ $(J \cdot m)$ | $1.87\text{x}10^{-64}$ | $3.22\times10^{-31}$ | $2.31\times10^{-28}$ | $2.5\times10^{-27}$ |
| Range *(m)* | infinity | $2\times10^{-18}$ | infinity | $1.5\times10^{-15}$ |

The potential energy associated with each force acting between two protons is characterized by both the strength of the interaction and the range over which the interaction takes place. In each case the strength is determined by a coupling constant, and the range is characterized by the mass of the exchanged particle. The potential energy, *U*, between two protons a distance *r* apart is written as

$$U = \frac{c^2}{r}\exp(-r/R),$$

where *R* is the range of the interaction, and $C^2$, is the strength of the interaction. In each case the interaction is due to the exchange of some particle whose mass determined the range of the interaction, $R = h/mc$. The exchanged particle is said to *mediate* the interaction.

A comparison between the coupling constants and ranges of the four forces acting between two protons. Although the graviton has not yet been observed, it is thought that there is an exchange particle associated with gravity and that eventually gravity will be described in a unified theory with the other three forces of nature.

The range of the gravitational and electromagnetic forces is infinite, while that of the strong and weak forces is very short. Also, the strength of the interaction depends on the separation between the two protons. Both gravitation and electromagnetism are of infinite range and their strengths decrease as the separation, *r*, increases–falling off as $1/r^2$. On the other hand, because the exchange particles for the strong and weak forces have a large mass, the force associated with them is zero outside of a short range.

Note that the strong force between two protons is a residual interaction. The quarks inside the nucleons interact through the exchange of gluons that carry a quantum number called "colour" (i.e., a "colour-neutral" object does not feel the strong force, like an electrically neutral object does not feel electromagnetism).

The theory of strong interactions among quarks and gluons is called Quantum Chromodynamics (QCD). A very successful model of describing this residual force is by the exchange of the π (pi) meson, or pion. This description, dating back to the work of Yukawa in the 1930s, works extremely well except at very short distances, and is generally used to describe most of the properties of complex nuclei. Exactly at what distance inside the nucleus, or correspondingly at what values of momentum transferred between the nucleons, the pion exchange model breaks down is the subject of much current research in nuclear physics. It is presently thought that this transition occurs somewhere around momentum transfers of 1 GeV/c.

## RADIATION IN THE ENVIRONMENT

Many forms of "radiation" are encountered in the natural environment and are produced by modern technology. Most of them have the potential for both beneficial and harmful effects. Even sunlight, the most essential radiation of all, can be harmful in excessive amounts. Most public attention is given to the category of radiation known as "ionizing radiation." This radiation can disrupt atoms, creating positive ions and negative electrons, and cause biological harm. Ionizing radiation includes x-rays, gamma rays, alpha particles, beta particles, neutrons, and the varieties of cosmic rays.

### Radiation Damage and its Study

All ionizing radiations, at sufficiently large exposures, can cause cancer. Many, in carefully controlled exposures, are also used for cancer therapy. Whether harmful or beneficial, exposures to ionizing radiation have been an inevitable part of the environment throughout the Earth's history. The nucleosynthesis processes that produced the elements created both stable and unstable nuclides. The unstable nuclides with very long half-lives, together with their radioactive progeny, constitute the natural radioactivity on Earth today. In addition, violent processes in the sun and elsewhere lead to the bombardment of the Earth by cosmic rays. Thus, radiation is an old and familiar, if unrecognized, pollutant.

However, human awareness of radioactivity and ionizing radiation has only a 100-year history starting with the discovery of x-rays and radioactivity. The first evidence that ionizing radiation could do harm came within months after the discovery of x-rays, when an early x-ray worker developed injuries to his skin. Serious efforts to understand and control radiation exposures started in the 1920s and greatly expanded during and after World War II.

Information on the effects of radiation comes from studies of exposed groups and individuals, from animal experiments, and from studies at the cellular and molecular level. It is now well established that ionizing radiation has both prompt and delayed effects. At very high radiation exposures, death will occur within several months or less. At moderate levels, radiation exposure increases the chance that an individual will develop cancer, with a time delay of ten or more years for most cancers. At low levels, the cancer risk decreases, but the relationship between cancer risk and the magnitude of the exposure is uncertain.

Other effects of radiation, in part inferred from animal experiments, include an increased risk of genetic defects and, for exposures of the fetus before birth, of mental retardation. In terms of frequency of occurrence and severity of effects, cancer is the most serious consequence and receives the greatest attention.

The importance of genetic effects has turned out to be much less than was originally expected. In the words of a 1993 NCRP report: "... the genetic risks have been found to be smaller and the cancer risks larger than were thought (in the 1950s)." Strikingly, no statistically significant genetic effects have been found among the extensively studied children of survivors of the Hiroshima-Nagasaki bombings. On the basis of animal experiments, however, one expects some genetic effects–even if so far not observable above the large background of "natural" defects. In addition, the Hiroshima-Nagasaki studies have shown an increased incidence of mental retardation among children who received large prenatal radiation doses, especially for exposures 8 to 15 weeks after conception.

**Units of Radioactivity and Dose**

The original unit for measuring the amount of radioactivity was the curie (Ci)–first defined to correspond to one gram of radium-226 and more recently defined as:

1 curie = $3.7 \times 10^{10}$ radioactive decays per second [exactly].

In the International System of Units (SI) the curie has been replaced by the becquerel (Bq), where

1 becquerel = 1 radioactive decay per second = $2.703 \times 10^{-11}$ Ci.

The magnitude of radiation exposures is specified in terms of the radiation dose. There are two important categories of dose:

The *absorbed dose,* sometimes also known as the *physical dose,* defined by the amount of energy deposited in a unit mass in human tissue or other media. The original unit is the *rad* [100 erg/g]; it is now being widely replaced by the SI unit, the gray (Gy) [1 J/kg], where 1 gray = 100 rad.

- The *biological* dose, sometimes also known as the *dose equivalent,* expressed in units of rem or, in the SI system, *sievert* (Sv). This dose reflects the fact that the biological damage caused by a particle depends not only on the total energy deposited but also on the rate of energy loss per unit distance traversed by the particle (or "linear energy transfer"). Alpha particles do much more damage per unit energy deposited than do electrons. This effect can be represented, in rough overall terms, by a *quality factor,* $Q$. Over a wide range of incident energies, $Q$ is taken to be 1.0 for electrons (and for x-rays and gamma rays, both of which produce electrons) and 20 for alpha particles. For neutrons, the adopted quality factor varies from 5 to 20, depending on neutron energy.

The biological impact is specified by the *dose equivalent* *H*, which is the product of the absorbed dose *D* and the quality factor *Q*: $H = Q\,D$.

The unit for the dose equivalent is the rem if the absorbed dose is in rads and the sievert (Sv) if the absorbed dose is in grays. Thus, 1 Sv = 100 rem. 1 rem is roughly the average dose received in 3 years of exposure to natural radiation. 1 Sv is at the bottom of the range of doses that, if received over a short period of time, are likely to cause noticeable symptoms of radiation sickness.

The dose equivalent is still not the whole story. If only part of the body is irradiated, the dose must be discounted with an appropriate weighting factor if it is to reflect overall risk. The discounted dose is termed the *effective dose equivalent* or just the *effective dose*, expressed in rems or sieverts.

## Radioactivity in the Natural Environment

The radioactive nuclei, or radionuclides, found naturally on Earth can be grouped into three series–headed by uranium-238, uranium-235, and thorium-232–plus several isolated beta-particle emitting nuclei, most prominently potassium-40 and rubidium-87.

**Table Half-lives and Average Abundances of Natural Radionuclides.**

| | $^{40}$K | $^{87}$Rb | $^{232}$Th | $^{238}$U |
|---|---|---|---|---|
| Half-life (billion years) | 1.277 | 47.5 | 14.05 | 4.468 |
| Upper continental crust | | | | |
| Elemental abundance (ppm) | 28000 | 112 | 10.7 | 2.8 |
| Activity (Bq/kg) | 870 | 102 | 43 | 35 |
| Activity (nCi/kg) | 23 | 2.7 | 1.2 | 0.9 |
| Activity (kCi/km$^3$) | 66 | 8 | 3.3 | 2.6 |
| Oceans | | | | |
| Elemental concentration (mg/liter) | 399 | 0.12 | $1\times10^{-7}$ | 0.0032 |
| Activity (Bq/liter) | 12 | 0.11 | $4\times10^{-7}$ | 0.040 |

| | | | | |
|---|---|---|---|---|
| Activity (nCi/liter) | 0.33 | 0.003 | $1x10^{-8}$ | 0.0011 |
| Ocean sediments | | | | |
| Elemental abundance (ppm) | 17000 | | 5.0 | 1.0 |
| Activity (Bq/kg) | 500 | | 20 | 12 |
| Activity (nCi/kg) | 14 | | 0.5 | 0.3 |
| Human body | | | | |
| Total activity (Bq) | 4000 | 600 | 0.08 | 0.4* |
| Total activity (nCi) | 100 | 16 | 0.002 | 0.01 |

In the human body the activity of $^{210}Pb$ and $^{210}Po$, both progeny of $^{238}U$, is much greater than that of $^{238}U$ itself.

The most interesting of the series is the uranium-238 series which decays via a chain containing 8 alpha decays and 6 beta decays to lead-206. This chain includes the longest-lived isotopes of radium and radon: radium-226 and radon-222, respectively. In each of the three chains the parent nucleus has a much greater lifetime than does any of the progeny. Therefore, a steady-state is established in which, for a given sample of material, each member of the series has the same activity–aside from deviations due to differences in chemical properties, which cause different elements to be transferred at different rates into or out of a given sample of material.

Including all the succeeding decays, the total activity in the thorium-232 and uranium-238 series is, very roughly, ten times the activity indicated for thorium-232 and uranium-238 alone. Thus, for each of the series, the total activity in the Earth's crust averages roughly 30,000 Ci/km$^3$. For both series together and including the contributions of potassium-40 and rubidium-87, the total activity in the crust averages about 100,000 Ci/km$^3$. There is also a considerable amount of radioactivity in the oceans, with potassium-40 dominant in the ocean itself and thorium-232 relatively more important in the ocean sediments. For the oceans as a whole ($1.4 \times 10^{21}$ liters), the total activity is about $4 \times 10^{11}$ Ci for potassium–40 and $1 \times 10^{9}$ Ci for uranium-238. Potassium–40 is also present in significant amounts in the human body, especially in muscle tissue.

In addition to these ancient radionuclides and their progeny, some radionuclides are being continually produced by cosmic rays. The most prominent of these is carbon-14, produced in the interaction of cosmic ray neutrons with nitrogen in the atmosphere.

**Table. Average Radiation doses.**

| Radiation source | Comments | Effective dose mSv/yr | mrem/yr |
|---|---|---|---|
| Natural sources | | | |
| indoor radon | due to seepage of $^{222}$Rn from ground | 2.0 | 200 |
| radionuclides in body | primarily $^{40}$K and $^{238}$U progeny | 0.39 | 39 |
| terrestrial radiation | due to gamma-ray emitters in ground | 0.28 | 28 |
| cosmic rays | roughly doubles for 2000 m gain in elevation | 0.27 | 27 |
| cosmogenic | especially $^{14}$C | 0.01 | 1 |
| total (rounded) | | 3.0 | 300 |
| Medical sources | | | |
| Diagnostic x-rays | excludes dental examinations | 0.39 | 39 |
| Medical treatments | radionuclides used in diagnosis (only) | 0.14 | 14 |
| total | | 0.53 | 53 |
| Other | | | |
| consumer products | primarily drinking water, building materials | 0.1 | 10 |
| occupational | averaged over population | 0.01 | 1 |
| nuclear fuel cycle | does not include potential reactor accidents | 0.0005 | 0.05 |
| Total (rounded) | | 3.6 | 360 |

*From *Ionizing Radiation Exposure of the Population of the United States*, NCRP Report No. 93 (National Council on Radiation Protection and Measurements, Washington DC, 1987).

## Typical Radiation Doses

The chief sources of radiation exposure in the United States, as tabulated by the NCRP. The largest single source of exposure is from radon, which is produced in the decay of radium-226 in the soil and enters a house through openings at the base. The "radon" dose arises mostly from the inhalation of the progeny of radon-222, and varies widely from house-to-house depending upon the radium content of the underlying soil, its porosity, and the house construction. The average effective dose of 2.0 mSv/yr (200 mrem/yr) corresponds to the average radon concentration, but there are more than one million homes with radon levels that are more than five times as great.

## Effects of Low Doses

Most of the radiation doses that are received by members of the public and by radiation workers–both routinely and in accidents–are what are commonly referred to as "low doses." There is no precise definition of "low" but it would include doses 10 mSv per year. The average radiation doses received by people in the U.S. are in the "low dose" region. It is obviously important to determine the effects of low radiation doses–or, more precisely, the effects of small additions to the unavoidable natural background dose.

However, despite much study, these effects are not known, being too small to see unambiguously. The most prominent assumption, accepted by most official bodies, is the so-called *linearity hypothesis,* according to which the cancer risk is directly proportional to the magnitude of the dose, down to zero dose. In applying this assumption a consensus estimate is that the risk to a "typical" individual of an eventual fatal cancer is 0.00005 per mSv (or 0.05 per Sv). Thus, if 100,000 people each receive an added dose of 1 mSv, then 5 additional cancer deaths are to be expected. At the same time, while adopting the linearity hypothesis as a prudent working assumption, many of the leading studies have also indicated the possibility that small increases in radiation dose do not create any additional cancer risk. This reflects the considerable disagreement that exists within the scientific community as to the validity of the linearity hypothesis.

### Effects of Large Doses

Radiation doses above 3 Gy (300 rad) can be fatal and doses above 6 Gy (600 rad) are almost certain to be fatal, with death occurring within several months (in shorter times at higher doses). [Note: Very high doses are commonly expressed in grays, because the standard quality factor is not appropriate. For gamma rays and electrons, 1 Gy corresponds to 1 Sv.] Above 1 Gy, radiation causes a complex of symptoms, including nausea and blood changes, known as radiation sickness. For doses below 1 Sv (100 rem), there is little likelihood of radiation sickness, and the main danger is an increased cancer risk. The most important data base and analyses are from the RERF studies of the Hiroshima and Nagasaki survivors. In these studies, the exposure and medical histories are analysed for an exposed group (50,113 people) and an unexposed, or minimally exposed, group (36,459 people). Through 1990, there have been 4,741 cancer fatalities in the exposed group, of which 454 are attributed to radiation exposure. There is a statistically significant excess for both solid cancer tumors and leukemia for doses above 0.2 Sv (20 rem). These data, in a succession of updated versions, have provided much of the information used in comprehensive studies of radiation effects.

### Nuclear Reactor Accidents

The accidents at the Three Mile Island (TMI) and Chernobyl nuclear reactors have triggered particularly intense concern about radiation hazards. The TMI accident, in Pennsylvania in 1979, resulted from a combination of deficient equipment and operator errors. Even though there was severe damage to the nuclear fuel within the reactor, very little radioactivity escaped into the outside environment. The effectiveness of the large concrete containment building that surrounded the reactor contributed to this relatively small release. Subsequent studies concluded that the maximum dose received by any member of the public was under 1 mSv (100 mrem). The collective off-site dose is estimated to have been about 20 person-Sv. Under the standard low-dose assumption, this corresponds to one eventual cancer fatality in the neighbouring population of 2 million people. (This population receives an annual collective dose of about 6000 person-Sv from natural sources.)

The 1986 Chernobyl accident was far more serious. It occurred in a reactor with an unsafe reactor design unique to the Soviet Union. The reactor had no effective containment, and there was a very large release of radionuclides to the environment. The accident led to the death within several months of 31 reactor personnel and firefighters–28 from a combination of radiation effects and burns from fire, 2 from other injuries, and one from a heart attack. A total of 237 workers was hospitalized for symptoms of radiation sickness, including the 28 who died. A 1996 summary reported an additional 14 deaths among the more severely exposed workers, but it is not clear that these deaths were all due to the prior exposure.

There is strong evidence of a substantial increase in thyroid cancers among children living in the vicinity. No other health effects from Chernobyl have been convincing established. However, it is too soon for them to have been fully manifested. Standard calculations of radiation effects predict that there will be a large number of excess cancer deaths among the so-called "liquidators," who were engaged in cleanup operations after the accident, as well as in the neighbouring population.

Considering impacts at greater distances, one early study estimated that the collective dose in the Northern Hemisphere over a 50-year period would be about 930,000 person-Sv. While there is substantial uncertainty in the dose estimate, there is even greater uncertainty as to the impact. If one accepts the linearity hypothesis and assumes 0.05 fatalities per Sv, this dose corresponds to 47,000 eventual cancer fatalities. About

29,000 of these fatalities would occur in Europe (outside the former Soviet Union) due to a cumulative collective dose of 580,000 Sv–an average individual *lifetime* dose of 1.2 mSv for 490 million people. Given these low average doses, any estimate of predicted deaths from Chernobyl is highly speculative. The deaths will not be identifiable, being masked by the 88,000,000 "normal" cancer fatalities expected in this region during the 50-year period.

Chapter 9

# Nuclear Reactions

Nuclear reactions and nuclear scattering are used to measure the properties of nuclei. Reactions that exchange energy or nucleons can be used to measure the energies of binding and excitation, quantum numbers of energy levels, and transition rates between levels. A particle accelerator produces a beam of high-velocity charged particles (electrons, protons, alphas, or "heavy ions"), which then strikes a target nucleus. Nuclear reactions can also be produced in nature by high-velocity particles from cosmic rays, for instance in the upper atmosphere or in space. Beams of neutrons can be obtained from nuclear reactors or as secondary products when a charged-particle beam knocks out weakly-bound neutrons from a target nucleus. Nuclear reactions can also be produced by beams of photons, mesons, muons, and neutrinos.

In order for a nuclear reaction to occur, the nucleons in the incident particle, or projectile, must interact with the nucleons in the target. Thus the energy must be high enough to overcome the natural electromagnetic repulsion between the protons. This energy "barrier" is called the Coulomb barrier. If the energy is below the barrier, the nuclei will bounce off each other. Early experiments by Rutherford used low-energy alpha particles from naturally radioactive material to bounce off target atoms and measure the size of the target nuclei.

When a collision occurs between the incident particle and a target nucleus, either the beam particle scatters elastically leaving the target nucleus in its ground state or the target nucleus is internally excited and subsequently decays by emitting radiation or nucleons. A nuclear reaction is described by identifying the incident particle, target nucleus, and reaction products. When a neutron strikes a nitrogen nucleus, $^{14}N$, to produce a proton, $^{1}H$, and an isotope of carbon, $^{14}C$, the reaction is written as

$$ {}_{0}^{1}n + {}_{1}^{14}N \rightarrow {}_{6}^{14}C + {}_{1}^{1}H $$

Sometimes the reaction is abbreviated as. A number of conservation conditions apply to any reaction equation:

- The mass number *A* and the charge *Z* must balance on each side of the reaction arrow. Thus in the example the sum of superscripts 1 + 14 on the left equals the 14 + 1 on the right to balance the *A's*. The 0 + 7 subscripts on the left equals the 6 + 1 on the right to balance the *Z's*.
- The total energy before the reaction must equal the total energy after the reaction. The total energy includes the particle kinetic energies plus the energy equivalent of the particle rest masses, $E = mc^2$.
- Linear momenta before and after the reaction must be equal. For two-particle final states this means that a measurement of one particle's momentum determines the other particle's momentum.
- Quantum rules govern the balancing of the angular momentum, parity, and isospin of the nuclear levels.

A specific reaction is studied by measuring the angles and kinetic energies of the reaction products (the kinematic variables). Particle and radiation detectors designed for the expected charge and energy of each product are arranged around the target.

The most important quantity of interest for a specific set of kinematic variables is the reaction cross section. The cross section is a measure of the probability for a particular reaction to occur. This quantity, $\sigma$, which has the dimension of area, is measured by the experimental ratio

$$\sigma = \frac{\text{number of reactions particles emitted}}{\begin{pmatrix}\text{number of beam} \\ \text{particles per unit area}\end{pmatrix}\begin{pmatrix}\text{number of target} \\ \text{nuclei within the beam}\end{pmatrix}}$$

The cross section can also be calculated from a mathematical model of the nucleus by applying the rules of quantum mechanics. Comparing the measured and calculated values of the cross sections for many reactions validates the assumptions of the nuclear model.

Table. Shows some of the many types of nuclear reactions and what they teach us about nuclei and nuclear energy.

**Table: Nuclear Reaction Types**

| Reaction | What is learned |
|---|---|
| Nucleon- nucleon scattering | Fundamental nuclear force |
| Elastic scattering of nuclei | Nuclear size and interaction potential |
| Inelastic scattering to excited states | Energy level location and quantum numbers |
| Inelastic scattering to the continuum | Giant resonances (vibrational modes) |
| Transfer and knockout reactions | Details of the Shell Model |
| Fusion reactions | Astrophysical processes |
| Fission reactions | Properties of Liquid-drop Model |
| Compound nucleus formation | Statistical properties of the nucleus |
| Multifragmentation | Phases of nuclear matter, Collective Model |
| Pion reactions | Investigation of the nuclear "glue" |
| Electron scattering | Quark structure of nuclei |

The elastic scattering cross sections of protons and neutrons on a proton target give the essential data to reconstruct the nucleon-nucleon. A complete theory of nuclear structure and dynamics must start with this elemental interaction.

The systematics of nuclear sizes, shapes, binding energies, and other nuclear properties are the data that nuclear models are challenged to explain. Most of this data was the result of elastic and inelastic scattering of electrons, protons, and neutrons from nuclei found in the chart of the nuclides.

At high enough excitation energies, a nucleus can undergo a series of normal modes of collective oscillations called giant resonances. The nucleus rings like a bell at distinct frequencies with all the nucleons participating and sharing the excitation energy.

Fusion reactions are the combining of two nuclei to form a more massive nucleus. Many fusion reactions release large amounts of energy. An example is the combining of two isotopes of hydrogen (tritium and deuterium) to form helium and a neutron plus a large amount of kinetic energy in the reaction products:

$$^{3}H + ^{2}H \rightarrow ^{4}He + n + 17.6 \text{ MeV}.$$

This reaction as a potential electric power source. Another example of fusion is the reaction set that powers the Sun and other low-mass stars:

$$^{1}H + ^{1}H \rightarrow ^{2}H + e+ + n\ e,$$

$$^{2}H + ^{1}H \rightarrow ^{3}He + \gamma,$$

$$^{3}He + ^{3}He \rightarrow ^{4}He + ^{1}H + ^{1}H.$$

The net energy output from this chain is 26.7 MeV for each helium-4 nucleus formed.

Neutron-induced fission of massive nuclei into two lower-mass nuclei plus neutrons is also an energy source for power generation.

Compound nucleus formation is a reaction in which two nuclei combine into a single excited nucleus; the excited nucleus lives for a relatively long time and "forgets" how it was formed. The decay from this state of excitation is by "evapouration" of nucleons from the heated liquid drop of the compound nucleus, by gamma decay, or by fission of the compound nucleus. The statistical nature of this process teaches us about the average properties of excited states of complex nuclei.

Multifragmentation reactions, in which high-energy nuclei collide with other nuclei, are a method of creating nuclear matter in unusual conditions of density and excitation energy. These states may be in a different phase from normal nuclei and be characteristic of the matter in the early universe.

The fundamental force between nucleons in nuclei is dominated by the exchange of $\pi$ mesons (pions). When these particles are created in high-energy proton reactions, they can be used to bombard nuclear targets. When the pion interacts with a nucleus, it forms a resonance with one of the bound nucleons. The resonance is shifted and broadened compared to the reaction on a free nucleon. These changes reflect the influence of the neighbouring nucleons.

When nucleons are flung at one another, they can mesh briefly. During the time they are one nucleus, the quarks in the nucleons can interact with one another as if they were free particles. As with Rutherford scattering, an investigation of the angle that a particle is scattered gives information about the conditions inside the nucleons.

Nuclear reactions and their interpretation are the main activity of most nuclear scientists. The continuing development of accelerators and detectors permit the refinement of nuclear data and models to benefit basic science and nuclear applications.

## ENERGY FROM NUCLEAR SCIENCE

The isotope $^{235}U$, with an abundance of only 0.7percent in natural uranium, is commonly used to produce electricity in nuclear fission reactors. This isotope has the distinctive and useful property of undergoing nuclear fission through interaction with thermal-energy neutrons (neutrons with average speeds of only a few km/s). The other main isotope of uranium, $^{238}$U, does not undergo nuclear fission with thermal neutrons, but it does capture neutrons to form the isotope $^{239}$Np which then decays to $^{239}$Pu. This isotope of plutonium undergoes nuclear fission with thermal neutrons with a higher probability than that of $^{235}$U. The energy released in the fission of $^{235}$U and $^{239}$Pu, mainly in the form of kinetic energy of the fission fragments, provides the heat to run the turbines that generate electricity at a nuclear fission power plant.

## NUCLEAR FISSION ENERGY

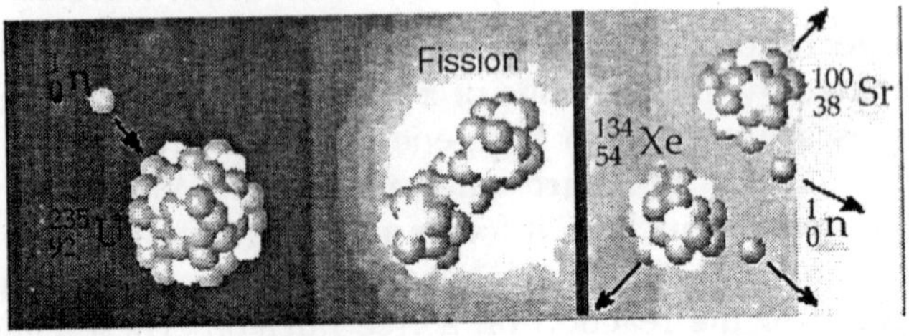


**Fig.** Fission of $^{235}$U after absorption of a thermal neutron.

The relevant nuclear reactions can be written as follows:

$^{235}$U + $^{1}$n Æ fission products + neutrons + energy (~200 MeV) (1)

$^{238}$U + $^{1}$n Æ $^{239}$U + gamma rays (2)

$^{239}$U Æ $^{239}$Np Æ $^{239}$Pu (a series of beta decays). (3)

An integer number of neutrons, either two, three or four, are emitted in the reactions leading to different pairs of fission products described in reaction (1). The average number is 2.43. Some of the neutrons in reaction (1) can be used to induce fission in another $^{235}$U nucleus, thus continuing a controlled, self-perpetuating nuclear chain reaction. Some fraction of the remaining neutrons from reaction (1) are utilized in reaction (2) to produce $^{239}$Pu. The rest are absorbed in other nuclei without further effect.

The isotope $^{232}Th$, although not fissionable with thermal neutrons, is a possible energy source because it absorbs thermal neutrons to produce long-lived $^{233}U$, which also undergoes thermal neutron fission with a high probability. Thus the "big three" readily fissionable nuclei are: $^{235}U$, $^{239}Pu$, and $^{233}U$.

A typical pressurized (or boiling) water nuclear reactor consists of a core of fissionable material ($UO_2$, enriched to 3.3percent to 4percent in $^{235}U$) in which the chain reaction takes place. The energy released in the fission process, which is primarily in the form of the kinetic energy of the fission fragments, heats the water. The water serves both as a neutron moderator (it slows down the fission neutrons to thermal energies), and as a heat transfer fluid. The chain reaction is controlled by rods of neutron-absorbing material inserted into the core. The thermal energy is removed from the core by the water to an external thermal-energy converter. In the pressurized water reactor (PWR), the thermal energy produces steam for the turbine through the use of a heat exchanger, whereas in a boiling water reactor (BWR), the steam is produced for direct use in the turbine.

Nuclear reactions liberate a large amount of energy compared to chemical reactions. One fission event results in the release of about 200 MeV of energy, or about $3.2 \times 10^{-11}$ watt-seconds. Thus, $3.1 \times 10^{10}$ fissions per second produce 1 W of thermal power. The fission of 1 g of uranium or plutonium per day liberates about 1 MW. This is the energy equivalent of 3 tons of coal or about 600 gallons of fuel oil per day, which when burned produces approximately 1/4 tonne of carbon dioxide. (A tonne, or metric ton, is 1000 kg.)

Nuclear reactors manufacture their own fuel, since they produce $^{239}Pu$ from $^{238}U$. With the total worldwide installed nuclear capacity of $3.4 \times 10^5$ $MW_e$ (megawatt electrical), one can estimate that more than 100 tonnes of $^{239}Pu$ are produced each year in reactors whose primary energy source is the fission of $^{235}U$. This $^{239}Pu$ can be reprocessed from used fuel rods and used to power other reactors.

It is actually possible to generate more $^{239}Pu$ than is used up in the reactor by surrounding the core with a uranium blanket and generating $^{239}Pu$ in this blanket. This is called a breeder reactor. A breeder reactor needs to be operated with fast neutrons, a so-called "fast breeder" reactor. In a fast-breeder reactor, water cannot be used as a coolant because it would moderate the neutrons. The smaller fission cross sections associated with the fast neutrons (as compared with thermal neutrons) leads to higher fuel concentrations in the core and higher power densities, which, in turn, create significant heat transfer problems. Liquid sodium metal may be used here as a coolant and heat-transfer fluid. Research on breeder reactors has essentially stopped in the United States because of concerns over nuclear proliferation since the plutonium bred in the reactor might be used for making weapons. Due to such concerns and the complexities of construction and operation, it is unlikely that breeder reactors will ever come into general operation within the next several decades, if ever.

In a yearly operating cycle of a typical (1000 $MW_e$) pressurized water reactor, the spent fuel contains about 25 tonnes of uranium and about 250 kg of $^{239}Pu$. Some 40percent of the energy produced in the course of a nuclear fuel cycle comes from $^{239}Pu$. Since about 20percent of the electricity generated in the United States comes from nuclear power plants, about twice as much electricity is generated from $^{239}Pu$ as is generated from oil-fired electrical generating plants.

Some 435 nuclear power plants operating around the world generate about 345,000 $MW_e$ of electricity in 32 countries, about one-sixth the world's electricity supply. Some countries depend vitally on the electricity generated by nuclear energy. France generates 76percent of its electricity from nuclear power plants; Belgium–56percent, South Korea–36percent, Switzerland–40percent, Sweden–47percent, Finland–30percent, Japan–33percent, and the United Kingdom–25percent. Bulgaria generates 46percent of its electricity from nuclear power, Hungary–42percent, and the Czech Republic and Slovakia combined–20percent. Although the United States is not a leader in percentage, it has the largest total electric output from nuclear power: 98,000 $MW_e$ from 105 plants, generating around 20percent of US electric power.

However, the United States, which led the world in early nuclear electric power development, was also first to be affected by its decline. Over the past two decades the US nuclear electric power industry has received no new domestic orders due to concerns over reactor safety, waste disposal, regulatory uncertainty, increased costs, and decreased electrical demand growth. These same pressures are now affecting nuclear power worldwide, although countries such as France and South Korea still have vigorous programmes. A rebirth may be imminent because of impending solutions to some of these problems and because of the problem of global warming due to the release of greenhouse gases in the combustion of fossil fuels.

The public has become suspicious of nuclear energy partly because of the elaborate methods used to address safety concerns. Most worrisome is the need for external electrical power to supply the pumps used in emergency core cooling systems in the types of reactors now used in the United States. In an accident, this external supply could possibly be disrupted leading to a release of radioactivity. New reactor designs use passive safety to address this problem. Passive safety features can be thought of as characteristics of a reactor that, without operator intervention, will tend to shut or cool a reactor down, keep it in a safe configuration, and prevent release of radioactivity.

These features fall into two broad categories–features that are designed to prevent accidents, and those that mitigate the effects of accidents. Many current problems arise from the huge scale of reactor construction projects. Typically, 1000 MW reactors have been constructed on site. The new types of reactors are smaller (e.g. 600 MW) and may be constructed in factories where uniformity and quality control can produce reactors and operating procedures less prone to failure. Examples of these new designs are two types of Advanced

Passive Light Water Reactors (APLWRs), the Liquid Metal Reactor (LMR), and the Modular High Temperature Gas-cooled Reactor (MHTGR). The APLWR is the most developed of these new reactor types. The AP-600 (Advanced Passive 600 $MW_e$), a pressurized water reactor. It runs at lower temperatures and with larger water inventories than current light water reactors, and has a passive emergency core-cooling system that utilizes gravity. It uses passive natural circulation for the removal of heat from the core.

Nuclear fission reactors, usually pressurized water reactors with energy conversion based on a steam-turbine cycle, have been used extensively to power ships. The United States has built and launched an impressive number of nuclear powered ships–a total of 155 attack and missile submarines, 9 guided missile cruisers (18 reactors), and 5 aircraft carriers (with a total of 24 reactors) and has operated 9 prototype reactors on land. Of this total, 22 submarines have been decommissioned or are non-operational. Shipboard reactors are constructed smaller than similar installations ashore, with special attention given to maintenance, protection from collision, and leakage. Although the primary use of nuclear power has been in submarines and other military vessels, a few non-military marine installations have been demonstrated. Examples include the Russian icebreaker Lenin and the United States demonstration merchant ship NS Savannah.

## Nuclear Fusion Power

Nuclear fusion reactors, if they can be made to work, promise virtually unlimited power for the indefinite future. This is because the fuel, isotopes of hydrogen, are essentially unlimited on Earth. Efforts to control the fusion process and harness it to produce power have been underway in the United States and abroad for more than forty years.

Nuclear fusion is the source of energy in the sun and stars where high temperatures and densities allow the positively-charged nuclei to get close enough to each other for the (attractive) nuclear force to overcome the (repulsive) electrical force and allow fusion to occur. The most promising fusion reaction,

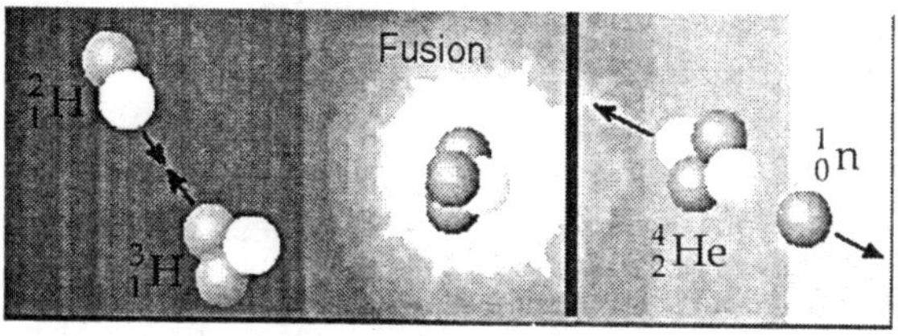


**Fig.** Nuclear Fusion

$^3H + ^2H$ Æ $^4He + n + 17.6$ MeV

involves the radioactive nuclide tritium ($^3H$), available from the nuclear production reaction

$^6Li + n$ Æ $^3H + ^4He$.

To produce energy using this reaction, both the magnetic confinement reactor with a high temperature plasma (a gas that has been completely ionized) and the inertial confinement reactor (which utilizes laser implosion technologies) have been investigated. Extremely high plasma temperatures are required in the magnetic confinement reactor and difficult laser implosion techniques are required for the inertial confinement reactor. Although significant progress has been made in these investigations, no working reactor that produces more energy than it consumes has been built. Unfortunately, the funding for continuing this work has declined, and the work is proceeding at a slower pace.

Although these types of reactors would not have the fission product waste disposal problem of fission reactors, fusion reactors generate large number of fast neutrons, leading to large quantities of radioactive byproducts.

Another approach to nuclear fusion–an approach that could lead to aneutronic power (power without neutrons) and non-radioactive nuclear energy–uses the concept of colliding-beam fusion (CBF). One aneutronic method features the $^2H + ^3He$ reaction leading to the products $^1H + ^4He$. However, this requires $^3He$ as fuel and terrestrial sources of this are limited. The Moon is a potential source of $^3He$ produced by cosmic-ray protons hitting the Moon directly and not being absorbed by an atmosphere as on Earth. Another potential approach for colliding beam fusion is the $^{11}B + ^{1}H$ reaction leading to the three $^{4}He$ nuclei. The energy release is in the form of charged particles whose kinetic energy can be converted to electricity with a very high efficiency. Current research predicts that this energy source has an extremely high degree of cleanness and efficiency. In all current energy sources, approximately two-thirds of the energy is lost in the form of waste heat or thermal pollution. In the CBF approach, there is virtually no waste. This design favours small size for the greatest efficiency (100 $MW_e$ or less), and would lead to either power plants with several reactors or decentralization of energy production.

## Chapter 10

# Nuclear Fusion

The deuterium-tritium *(D-T) fusion reaction is considered the m*ost promising for producing sustainable fusion power. From the top: 1. the Deuterium and Tritium nuclei are accelerated towards each other at thermonuclear speeds/ temperatures; 2. they combine to create an unstable helium-5 nucleus; 3. the He-5 nucleus decays, resulting in the ejection of a neutron and repulsion of the He-4 nucleus, both with high energies. For the generation of electrical power by fusion.

In physics and nuclear chemistry, nuclear fusion is the process by which multiple atomic particles join together to form a heavier nucleus. It is accompanied by the release or absorption of energy. Iron and nickel nuclei have the largest binding energies per nucleon of all nuclei and therefore are the most stable. The fusion of two nuclei lighter than iron or nickel generally releases energy while the fusion of nuclei heavier than iron or nickel absorbs energy; vice-versa for the reverse process, nuclear fission.

Nuclear fusion occurs naturally in stars. Artificial fusion in human enterprises has also been achieved, although not yet completely controlled. Building upon the nuclear transmutation experiments of Ernest Rutherford done a few years earlier, fusion of light nuclei (hydrogen isotopes) was first observed by Mark Oliphant in 1932, and the steps of the main cycle of nuclear fusion in stars were subsequently worked out by Hans Bethe throughout the remainder of that decade. Research into fusion for military purposes began in the early

1940s, as part of the Manhattan Project, but was not successful until 1952. Research into controlled fusion for civilian purposes began in the 1950s, and continues to this day.

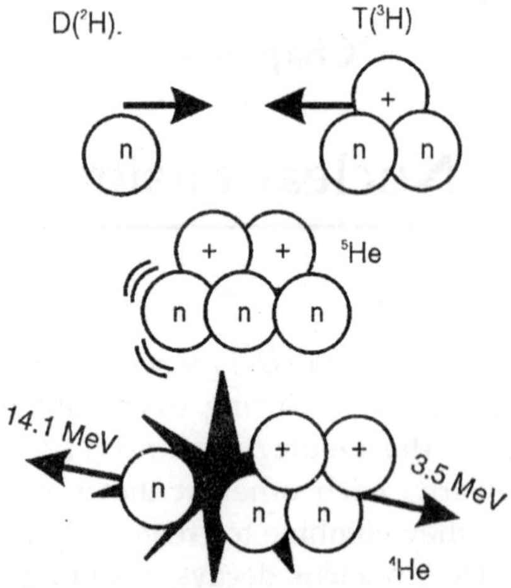


**Fig.** Fusion reaction

Fusion reactions power the stars and produce all but the lightest elements in a process called nucleosynthesis. Whereas the fusion of light elements in the stars releases energy, production of the heaviest elements absorbs energy, so it can only take place in the extremely high-energy conditions of supernova explosions.

When the fusion reaction is a sustained uncontrolled chain, it can result in a thermonuclear explosion, such as that generated by a hydrogen bomb. Reactions which are not self-sustaining can still release considerable energy, as well as large numbers of neutrons.

Research into controlled fusion, with the aim of producing fusion power for the production of electricity, has been conducted for over 50 years. It has been accompanied by extreme scientific and technological difficulties, and as of yet has not been successful in producing workable designs. As of the present, the only self-sustaining fusion reactions produced by humans have been produced in hydrogen bombs, where the extreme power of a fission bomb is necessary to begin the process. While some plans have been put forth to attempt to use the explosions of hydrogen bombs to generate electricity, none of these have ever moved far past the design stage.

It takes considerable energy to force nuclei to fuse, even those of the lightest element, hydrogen. This is because all nuclei have a positive charge (due to their protons), and as like charges repel, nuclei strongly resist being put too close together. Accelerated to high speeds (that is, heated to thermonuclear temperatures), however, they can overcome this electromagnetic repulsion and get close enough for the strong nuclear force to be active, achieving fusion. The fusion of lighter nuclei, creating a heavier nucleus and a free neutron, will generally release more energy than it took to force them together-an exothermic process that can produce self-sustaining reactions.

The energy released in most nuclear reactions is much larger than that in chemical reactions, because the binding energy that holds a nucleus together is far greater than the energy that holds electrons to a nucleus. The ionization energy gained by adding an electron to a hydrogen nucleus is 13.6 electron volts-less than one-millionth of the 17 MeV released in the D-T (deuterium-tritium) reaction shown to the top right. Fusion reactions have an energy density many times greater than nuclear fission-that is, per unit of mass the reactions produce far greater energies, even though *individual* fission reactions are generally much more energetic than *individual* fusion reactions-which are themselves millions of times more energetic than chemical reactions. Only the direct conversion of mass into energy, such as with collision of matter and antimatter, is more energetic per unit of mass than nuclear fusion.

## Requirements

A substantial energy barrier must be overcome before fusion can occur. At large distances two naked nuclei repel one another because of the repulsive electrostatic force between their positively charged protons. If two nuclei can be brought close enough together, however, the electrostatic repulsion can be overcome by the nuclear force which is stronger at close distances.

When a nucleon such as a proton or neutron is added to a nucleus, the nuclear force attracts it to other nucleons, but primarily to its immediate neighbours due to the short range of the force. The nucleons in the interior of a nucleus have more neighbouring nucleons than those on the surface. Since smaller nuclei have a larger surface area-to-volume ratio, the binding energy per nucleon due to the strong force generally increases with the size of the nucleus but approaches a limiting value corresponding to that of a fully surrounded nucleon.

The electrostatic force, on the other hand, is an inverse-square force, so a proton added to a nucleus will feel an electrostatic repulsion from *all* the other protons in the nucleus. The electrostatic energy per nucleon due to the electrostatic force thus increases without limit as nuclei get larger.

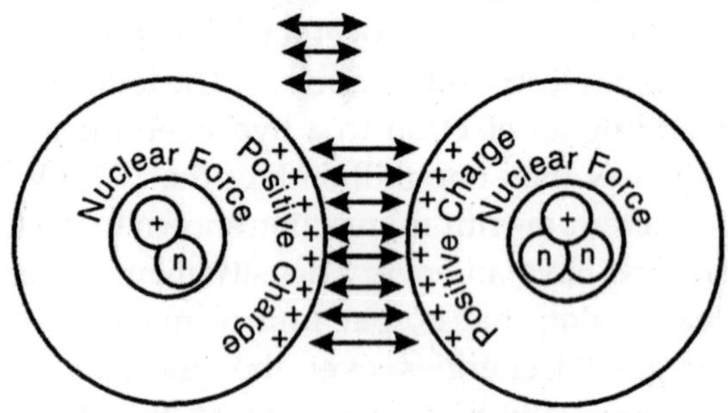


**Fig.** Electrostatic force of nuclei

The electrostatic force caused by positively charged nuclei is very strong over long distances, but at short distances the nuclear force is stronger. As such, the main technical difficulty for fusion is getting the nuclei close enough to fuse. Distances not to scale.

The net result of these opposing forces is that the binding energy per nucleon generally increases with increasing size, up to the elements iron and nickel, and then decreases for heavier nuclei. Eventually, the binding energy becomes negative and very heavy nuclei are not stable. The four most tightly bound nuclei, in decreasing order of binding energy, are $^{62}Ni$, $^{58}Fe$, $^{56}Fe$, and $^{60}Ni$. Even though the nickel isotope, $^{62}Ni$, is more stable, the iron isotope $^{56}Fe$ is an order of magnitude more common. This is due to a greater disintegration rate for $^{62}Ni$ in the interior of stars driven by photon absorption.

A notable exception to this general trend is the helium-4 nucleus, whose binding energy is higher than that of lithium, the next heavier element. The Pauli exclusion principle provides an explanation for this exceptional behaviour — it says that because protons and neutrons are fermions, they cannot exist in exactly the same state. Each proton or neutron energy state in a nucleus can accommodate both a spin up particle and a spin down particle. Helium-4 has an anomalously large binding energy because its nucleus consists of two protons and two neutrons; so all four of its nucleons can be in the ground state. Any additional nucleons would have to go into higher energy states.

The situation is similar if two nuclei are brought together. As they approach each other, all the protons in one nucleus repel all the protons in the other. Not until the two nuclei actually come in contact can the strong nuclear force take over. Consequently, even when the final energy state is lower, there is a large energy barrier that must first be overcome. It is called the Coulomb barrier.

The Coulomb barrier is smallest for isotopes of hydrogen—they contain only a single positive charge in the nucleus. A bi-proton is not stable, so neutrons must also be involved, ideally in such a way that a helium nucleus, with its extremely tight binding, is one of the products.

Using deuterium-tritium fuel, the resulting energy barrier is about 0.01 MeV. In comparison, the energy needed to remove an electron from hydrogen is 13.6 eV, about 750 times less energy. The (intermediate) result of the fusion is an unstable $^{5}He$ nucleus, which immediately ejects a neutron with 14.1 MeV. The recoil energy of the remaining $^{4}He$ nucleus is 3.5 MeV, so the total energy liberated is 17.6 MeV. This is many times more than what was needed to overcome the energy barrier.

If the energy to initiate the reaction comes from accelerating one of the nuclei, the process is called *beam-target* fusion; if both nuclei are accelerated, it is *beam-beam* fusion. If the nuclei are part of a plasma near thermal equilibrium, one speaks of *thermonuclear* fusion. Temperature is a measure of the average kinetic energy of particles, so by heating the nuclei they will gain energy and eventually have enough to overcome this 0.01 MeV. Converting the units between electronvolts and kelvins shows that the barrier would be overcome at a temperature in excess of 120 million kelvins, obviously a very high temperature.

There are two effects that lower the actual temperature needed. One is the fact that temperature is the *average* kinetic energy, implying that some nuclei at this temperature would actually have much higher energy than 0.01 MeV, while others would be much lower. It is the nuclei in the high-energy tail of the velocity distribution that account for most of the fusion reactions. The other effect is quantum tunneling. The nuclei do not actually have to have enough energy to overcome the Coulomb barrier completely. If they have nearly enough energy, they can tunnel through the remaining barrier. For this reason fuel at lower temperatures will still undergo fusion events, at a lower rate.

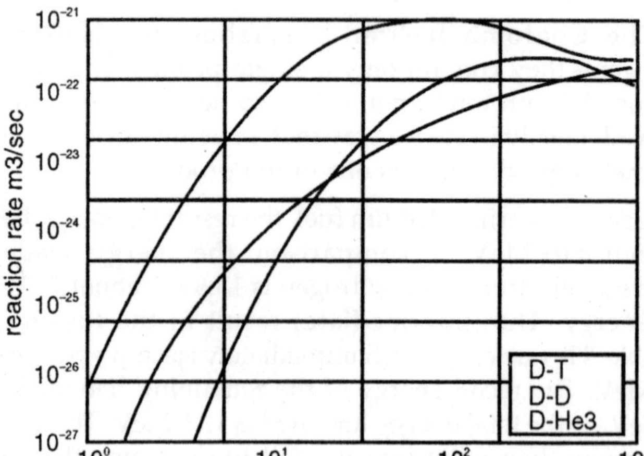

The fusion reaction rate increases rapidly with temperature until it maximizes and then gradually drops off. The DT rate peaks at a lower temperature (about 70 keV, or 800 million kelvins) and at a higher value than other reactions commonly considered for fusion energy.

The reaction cross section $\sigma$ is a measure of the probability of a fusion reaction as a function of the relative velocity of the two reactant nuclei. If the reactants have a distribution of velocities, e.g. a thermal distribution with thermonuclear fusion, then it is useful to perform an average over the distributions of the product of cross section and velocity. The reaction rate (fusions per volume per time) is $<\sigma v>$ times the product of the reactant number densities:

$$f = n_1 n_2 \langle \sigma v \rangle$$

If a species of nuclei is reacting with itself, such as the DD reaction, then the product $n_1 n_2$ must be replaced by $(1 / 2)n^2$.

$\langle \sigma v \rangle$ increases from virtually zero at room temperatures up to meaningful magnitudes at temperatures of 10 – 100 keV. At these temperatures, well above typical ionization energies (13.6 eV in the hydrogen case), the fusion reactants exist in a plasma state.

The significance of $\langle \sigma v \rangle$ as a function of temperature in a device with a particular energy confinement time is found by considering the Lawson criterion.

## FUEL CONFINEMENT METHODS

### Gravitational

One force capable of confining the fuel well enough to satisfy the Lawson criterion is gravity. The mass needed, however, is so great that gravitational confinement is only found in stars (the smallest of which are brown dwarfs). Even if the more reactive fuel deuterium were used, a mass greater than that of the planet Jupiter would be needed.

### Magnetic

The charged ions of fusion fuel follow spiral orbits around magnetic field lines, and the fuel is therefore trapped along the field lines. A variety of magnetic configurations exist, including the toroidal geometries of tokamaks and stellarators and open ended mirror confinement systems.

### Inertial

A third confinement principle is to apply a rapid pulse of energy to a large part of the surface of a pellet of fusion fuel, causing it to simultaneously "implode" and heat to very high pressure and temperature. If the fuel is dense enough and hot enough, the fusion reaction rate will be high enough to burn a significant fraction of the fuel before it has dissipated. To achieve these extreme conditions, the initially cold fuel must be explosively compressed. Inertial confinement is used in the hydrogen bomb, where the driver is x-rays created by a fission bomb. Inertial confinement is also attempted in "controlled" nuclear fusion, where the driver is a laser, ion, or electron beam, or a Z-pinch.

Some other confinement principles have been investigated, such as muon-catalyzed fusion, the Farnsworth-Hirsch fusor and Polywell (inertial electrostatic confinement), and bubble fusion.

## PRODUCTION METHODS

A variety of methods are known to effect nuclear fusion. Some are "cold" in the strict sense that no part of the material is hot (except for the reaction products), some are "cold" in the limited sense that the bulk of the material is at a relatively low temperature and pressure but the reactants are not, and some are "hot" fusion methods that create macroscopic regions of very high temperature and pressure.

### Locally Cold Fusion

- Muon-catalyzed fusion is a well-established and reproducible fusion process that occurs at ordinary temperatures. It was studied in detail by Steven Jones in the early 1980s. It has not been reported to produce net energy. Net energy production from this reaction is not believed to be possible because of the energy required to create muons, their 2.2 μs half-life, and the chance that a muon will bind to the new alpha particle and thus stop catalyzing fusion.

- "Cold fusion" also refers to a simple method using electrodes in heavy water, which has not been verified to be possible.

## Generally Cold, locally Hot Fusion

- Accelerator based light-ion fusion. Using particle accelerators it is possible to achieve particle kinetic energies sufficient to induce many light ion fusion reactions. Of particular relevance into this discussion are devices referred to as sealed-tube neutron generators. These small devices are miniature particle accelerators filled with deuterium and tritium gas in an arrangement which allows ions of these nuclei to be accelerated against hydride targets, also containing deuterium and tritium, where fusion takes place. Hundreds of neutron generators are produced annually for use in the petroleum industry where they are used in measurement equipment for locating and mapping oil reserves. Despite periodic reports in the popular press by scientists claiming to have invented "table-top" fusion machines, neutron generators have been around for half a century. The sizes of these devices vary but the smallest instruments are often packaged in sizes smaller than a loaf of bread. These devices do not produce a net power output.
- In sonoluminescence, acoustic shock waves create temporary bubbles that collapse shortly after creation, producing very high temperatures and pressures. In 2002, Rusi P. Taleyarkhan reported the possibility that bubble fusion occurs in those collapsing bubbles (aka sonofusion). As of 2005,

experiments to determine whether fusion is occurring give conflicting results. If fusion is occurring, it is because the local temperature and pressure are sufficiently high to produce hot fusion. In an episode of Horizon, on BBC television, it was conclusively shown that, although temperatures were reached which could initiate fusion on a large scale, no fusion was occurring, and inaccuracies in the measuring system were the cause of anomalous results.

- The Farnsworth-Hirsch Fusor is a tabletop device in which fusion occurs. This fusion comes from high effective temperatures produced by electrostatic acceleration of ions. The device can be built inexpensively, but it too is unable to produce a net power output.
- Antimatter-initialized fusion uses small amounts of antimatter to trigger a tiny fusion explosion. This has been studied primarily in the context of making nuclear pulse propulsion feasible. This is not near becoming a practical power source, due to the cost of manufacturing antimatter alone.
- Pyroelectric fusion was reported in April 2005 by a team at UCLA. The scientists used a pyroelectric crystal heated from –34 to 7°C (–30 to 45°F), combined with a tungsten needle to produce an electric field of about 25 gigavolts per meter to ionize and accelerate deuterium nuclei into an erbium deuteride target. Though the energy of the deuterium ions generated by the crystal has not been directly measured, the authors used 100 keV (a temperature of about $10^9$ K) as an estimate in their modeling..At these energy levels, two deuterium nuclei can fuse together to produce a helium-3 nucleus, a 2.45 MeV neutron and bremsstrahlung. Although it makes a useful neutron generator, the apparatus is not intended for power generation since it requires far more energy than it produces.

## Hot Fusion

In "standard" "hot" fusion, the fuel reaches tremendous temperature and pressure inside a fusion reactor or nuclear weapon.

The methods in the second group are examples of non-equilibrium systems, in which very high temperatures and pressures are produced in a relatively small region adjacent to material of much lower temperature. In his doctoral thesis for MIT, Todd Rider did a theoretical study of all quasineutral, isotropic, non-equilibrium fusion systems. He demonstrated that all such systems will leak energy at a rapid rate due to bremsstrahlung radiation produced when electrons in the plasma hit other electrons or ions at a cooler temperature and suddenly decelerate. The problem is not as pronounced in a hot plasma because the range of temperatures, and thus the magnitude of the deceleration, is much lower. Note that Rider's work does not apply to non-neutral and/or anisotropic non-equilibrium plasmas.

## IMPORTANT REACTIONS

### Astrophysical Reaction Chains

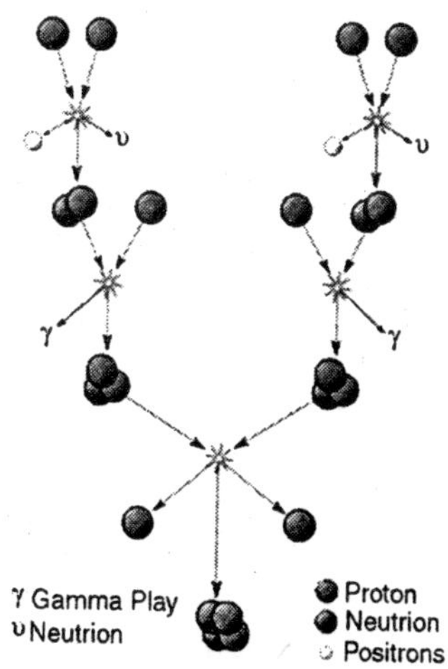


**Fig.** Chain reaction

The proton-proton chain dominates in stars the size of the Sun or smaller.

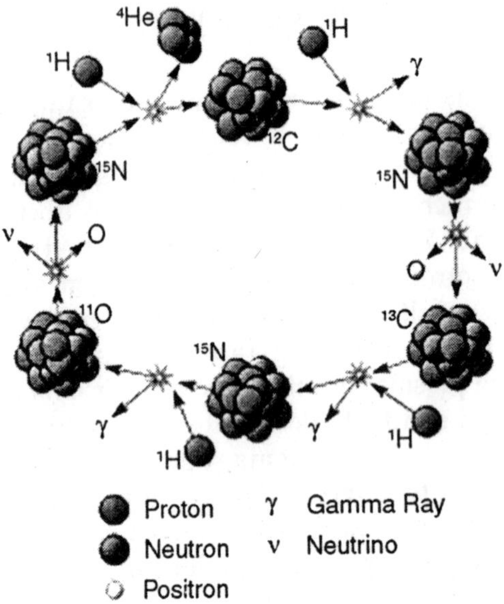


**Fig.** The CNO cycle dominates in stars heavier than the Sun.

The most important fusion process in nature is that which powers the stars. The net result is the fusion of four protons into one alpha particle, with the release of two positrons, two neutrinos (which changes two of the protons into neutrons), and energy, but several individual reactions are involved, depending on the mass of the star. For stars the size of the sun or smaller, the proton-proton chain dominates. In heavier stars, the CNO cycle is more important. Both types of processes are responsible for the creation of new elements as part of stellar nucleosynthesis.

At the temperatures and densities in stellar cores the rates of fusion reactions are notoriously slow. At solar core temperature ($T \approx 15$ MK) and density (120 g/cm$^3$), the energy release rate is only 276 mW/cm$^3$—about a quarter of the volumetric rate at which a resting human body generates heat.

Thus, reproduction of stellar core conditions in a lab for nuclear fusion power production is completely impractical. Because nuclear reaction rates strongly depend on temperature (exp(–*E*/*kT*)), then in order to achieve reasonable rates of energy production in terrestrial fusion reactors 10–100 times higher temperatures (compared to stellar interiors) are required $T \approx 0.1–1.0$ GK.

## Criteria and Candidates for Terrestrial Reactions

In man-made fusion, the primary fuel is not constrained to be protons and higher temperatures can be used, so reactions with larger cross-sections are chosen. This implies a lower Lawson criterion, and therefore less startup effort. Another concern is the production of neutrons, which activate the reactor structure radiologically, but also have the advantages of allowing volumetric extraction of the fusion energy and tritium breeding. Reactions that release no neutrons are referred to as *aneutronic.*

In order to be useful as a source of energy, a fusion reaction must satisfy several criteria. It must

- Be exothermic: This may be obvious, but it limits the reactants to the low Z (number of protons) side of the curve of binding energy. It also makes helium He-4 the most common product because of its extraordinarily tight binding, although He-3 and H-3 also show up;
- Involve low Z nuclei: This is because the electrostatic repulsion must be overcome before the nuclei are close enough to fuse;
- Have two reactants: At anything less than stellar densities, three body collisions are too improbable. It should be noted that in inertial confinement, both stellar densities and temperatures are exceeded to compensate for the shortcomings of the third parameter of the Lawson criterion, ICF's very short confinement time;
- Have two or more products: This allows simultaneous conservation of energy and momentum without relying on the electromagnetic force;

- Conserve both protons and neutrons: The cross sections for the weak interaction are too small.

For reactions with two products, the energy is divided between them in inverse proportion to their masses, as shown. In most reactions with three products, the distribution of energy varies. For reactions that can result in more than one set of products, the branching ratios are given.

Some reaction candidates can be eliminated at once. The D-$^6$Li reaction has no advantage compared to p-$^{11}$B because it is roughly as difficult to burn but produces substantially more neutrons through D-D side reactions. There is also a p-$^7$Li reaction, but the cross section is far too low, except possibly when $T_i > 1$ MeV, but at such high temperatures an endothermic, direct neutron-producing reaction also becomes very significant. Finally there is also a p-$^9$Be reaction, which is not only difficult to burn, but $^9$Be can be easily induced to split into two alphas and a neutron.

In addition to the fusion reactions, the following reactions with neutrons are important in order to "breed" tritium in "dry" fusion bombs and some proposed fusion reactors:

$$n + {}^6\text{Li} \rightarrow T + {}^4\text{He}$$

$$n + {}^7\text{Li} \rightarrow T + {}^4\text{He} + n$$

To evaluate the usefulness of these reactions, in addition to the reactants, the products, and the energy released, one needs to know something about the cross section. Any given fusion device will have a maximum plasma pressure that it can sustain, and an economical device will always operate near this maximum. Given this pressure, the largest fusion output is obtained when the temperature is chosen so that $\langle \sigma v \rangle / T^2$ is a maximum.

This is also the temperature at which the value of the triple product $nT\hat{o}$ required for ignition is a minimum, since that required value is inversely proportional to $\langle \sigma v \rangle / T^2$. (A plasma is "ignited" if the fusion reactions produce enough power to maintain the temperature without external heating.) This optimum temperature and the value of $\langle \sigma v \rangle / T^2$ at that temperature is given for a few of these reactions in the following table.

| **fuel** | $T$ **[keV]** | $\langle \sigma v \rangle / T^2$ **[m³/s/keV²]** |
|---|---|---|
| D-T | 13.6 | $1.24 \times 10^{-24}$ |
| D-D | 15 | $1.28 \times 10^{-26}$ |
| D-$^3$He | 58 | $2.24 \times 10^{-26}$ |
| p-$^6$Li | 66 | $1.46 \times 10^{-27}$ |
| p-$^{11}$B | 123 | $3.01 \times 10^{-27}$ |

Note that many of the reactions form chains. For instance, a reactor fueled with T and $^3$He will create some D, which is then possible to use in the D + $^3$He reaction if the energies are "right". An elegant idea is to combine the reactions (8) and (9). The $^3$He from reaction (8) can react with $^6$Li in reaction (9) before completely thermalizing. This produces an energetic proton which in turn undergoes reaction (8) before thermalizing. A detailed analysis shows that this idea will not really work well, but it is a good example of a case where the usual assumption of a Maxwellian plasma is not appropriate.

## Neutronicity, Confinement Requirement, and Power Density

The only fusion reactions thus far produced by humans to achieve ignition are those which have been created in hydrogen bombs; the first of which, Ivy Mike, is shown here.

Any of the reactions above can in principle be the basis of fusion power production. In addition to the temperature and cross section discussed above, we must consider the total energy of the fusion products $E_{fus}$, the energy of the charged fusion products $E_{ch}$, and the atomic number $Z$ of the non-hydrogenic reactant.

Specification of the D-D reaction entails some difficulties, though. To begin with, one must average over the two branches (2) and (3). More difficult is to decide how to treat the T and $^3$He products. T burns so well in a deuterium plasma that it is almost impossible to extract from the plasma. The D-$^3$He reaction is optimized at a much higher temperature, so the burnup at the optimum D-D temperature may be low, so it seems reasonable to assume the T but not the $^3$He gets burned up and adds its energy to the net reaction. Thus we will count the DD fusion energy as $E_{fus}$ = (4.03+17.6+3.27)/2 = 12.5 MeV and the energy in charged particles as $E_{ch}$ = (4.03+3.5+0.82)/2 = 4.2 MeV.

Another unique aspect of the D-D reaction is that there is only one reactant, which must be taken into account when calculating the reaction rate.

With this choice, we tabulate parameters for four of the most important reactions.

| fuel | Z | $E_{fus}$ [MeV] | $E_{ch}$ [MeV] | neutronicity |
|---|---|---|---|---|
| D-T | 1 | 17.6 | 3.5 | 0.80 |
| D-D | 1 | 12.5 | 4.2 | 0.66 |
| D-$^3$He | 2 | 18.3 | 18.3 | ~0.05 |
| p-$^{11}$B | 5 | 8.7 | 8.7 | ~0.001 |

The last column is the neutronicity of the reaction, the fraction of the fusion energy released as neutrons. This is an important indicator of the magnitude of the problems associated with neutrons like radiation damage, biological shielding, remote handling, and safety. For the first two reactions it is calculated as $(E_{fus}-E_{ch})/E_{fus}$. For the last two reactions, where this calculation would give zero, the values quoted are rough estimates based on side reactions that produce neutrons in a plasma in thermal equilibrium.

Of course, the reactants should also be mixed in the optimal proportions. This is the case when each reactant ion plus its associated electrons accounts for half the pressure. Assuming that the total pressure is fixed, this means that density of the non-hydrogenic ion is smaller than that of the hydrogenic ion by a factor 2/(Z+1). Therefore the rate for these reactions is reduced by the same factor, on top of any differences in the values of $\langle\sigma v\rangle > T^2$. On the other hand, because the D-D reaction has only one reactant, the rate is twice as high as if the fuel were divided between two hydrogenic species.

Thus there is a "penalty" of (2/(Z+1)) for non-hydrogenic fuels arising from the fact that they require more electrons, which take up pressure without participating in the fusion reaction. (It is usually a good assumption that the electron temperature will be nearly equal to the ion temperature. Some authors, however discuss the possibility that the electrons could be maintained substantially colder than the ions. In such a case, known as a "hot ion mode", the "penalty" would not apply.) There is at the same time a "bonus" of a factor 2 for D-D due to the fact that each ion can react with any of the other ions, not just a fraction of them.

We can now compare these reactions in the following table.

| **fuel** | $\langle\sigma v\rangle/T^2$ | **penalty/ bonus** | **reactivity** | **Lawson criterion** | **power density** |
|---|---|---|---|---|---|
| D-T | $1.24\times10^{-24}$ | 1 | 1 | 1 | 1 |
| D-D | $1.28\times10^{-26}$ | 2 | 48 | 30 | 68 |
| D-$^3$He | $2.24\times10^{-26}$ | 2/3 | 83 | 16 | 80 |
| p-$^{11}$B | $3.01\times10^{-27}$ | 1/3 | 1240 | 500 | 2500 |

The maximum value of $\langle\sigma v\rangle > T^2$ is taken from a previous table. The "penalty/bonus" factor is that related to a non-hydrogenic reactant or a single-species reaction. The values in the column "reactivity" are found by dividing $1.24 \times 10^{-24}$ by the product of the second and third columns. It indicates the factor by which the other reactions occur more slowly than the D-T reaction under comparable conditions. The column "Lawson criterion" weights these results with $E_{ch}$ and gives an indication of how much more difficult it is to achieve ignition with these reactions, relative to the difficulty for the D-T reaction. The last column is labeled "power density" and weights the practical reactivity with

$E_{fus}$. It indicates how much lower the fusion power density of the other reactions is compared to the D-T reaction and can be considered a measure of the economic potential.

### Bremsstrahlung Losses in Quasineutral, Isotropic Plasmas

The ions undergoing fusion in many systems will essentially never occur alone but will be mixed with electrons that in aggregate neutralize the ions' bulk electrical charge and form a plasma. The electrons will generally have a temperature comparable to or greater than that of the ions, so they will collide with the ions and emit x-ray radiation of 10-30 keV energy (Bremsstrahlung). The Sun and stars are opaque to x-rays, but essentially any terrestrial fusion reactor will be optically thin for x-rays of this energy range. x-rays are difficult to reflect but they are effectively absorbed (and converted into heat) in less than mm thickness of stainless steel (which is part of a reactor's shield). The ratio of fusion power produced to x-ray radiation lost to walls is an important figure of merit. This ratio is generally maximized at a much higher temperature than that which maximizes the power density. The following table shows the rough optimum temperature and the power ratio at that temperature for several reactions.

| fuel | $T_i$ (keV) | $P_{fusion}/P_{Bremsstrahlung}$ |
|---|---|---|
| D-T | 50 | 140 |
| D-D | 500 | 2.9 |
| D-$^3$He | 100 | 5.3 |
| $^3$He-$^3$He | 1000 | 0.72 |
| p-$^6$Li | 800 | 0.21 |
| p-$^{11}$B | 300 | 0.57 |

The actual ratios of fusion to Bremsstrahlung power will likely be significantly lower for several reasons. For one, the calculation assumes that the energy of the fusion products is transmitted completely to the fuel ions, which then lose energy to the electrons by collisions, which in turn lose energy by Bremsstrahlung. However because the fusion products move much faster than the fuel ions, they will give up a significant fraction of their energy directly to the electrons. Secondly, the plasma is assumed to be composed purely of fuel ions. In practice, there will be a significant proportion of impurity ions, which will lower the ratio. In particular, the fusion products themselves *must* remain in the plasma until they have given up their energy, and *will* remain some time after that in any proposed confinement scheme. Finally, all channels of energy loss other than Bremsstrahlung have been neglected. The last two factors are related. On theoretical and experimental grounds, particle and energy confinement seem to be closely related. In a confinement scheme that does a good job of retaining energy, fusion products will build up. If the fusion products are efficiently ejected, then energy confinement will be poor, too.

The temperatures maximizing the fusion power compared to the Bremsstrahlung are in every case higher than the temperature that maximizes the power density and minimizes the required value of the fusion triple product. This will not change the optimum operating point for D-T very much because the Bremsstrahlung fraction is low, but it will push the other fuels into regimes where the power density relative to D-T is even lower and the required confinement even more difficult to achieve. For D-D and D-$^3$He, Bremsstrahlung losses will be a serious, possibly prohibitive problem. For $^3$He-$^3$He, p-$^6$Li and p-$^{11}$B the Bremsstrahlung losses appear to make a fusion reactor using these fuels with a quasineutral, anisotropic plasma impossible. This limitation does not apply to non-neutral and anisotropic plasmas; however, these have their own challenges to contend with.

## Advantages and Disadvantages

### *Advantages:*

- The fuel for fusion reactions are readily available. Deuterium and Tritium are virtually inexhaustible.
- Unlike the burning of coal or other fossil fuels, fusion does not emit harmful toxins into the atmosphere. The combustion of most fossil fuels involves some form of the reaction

$$C + O_2(g) \rightarrow CO_2(g) + heat$$

- The carbon dioxide ($CO_2(g)$) emitted by this reaction contributes to the global warning/so-called "Greenhouse Effect" that we've all heard so much about. Fusion, however, produces only helium, a gas that is already in abundance in the atmosphere and will not contribute to global warming.
- A major concern with the use of fission power is the issue of nuclear waste, a dangerous material that can both directly injure people and be manufactured into weapons. Fusion has no such problems with dangerous by-products.

***Disadvantages***

- Scientists have not yet been able to contain a fusion reaction long enough for there to be a net energy gain.
- Many countries are phasing out fusion research because of the failure to reach a breakthrough.

## THEORY

The theory behind fusion is actually relatively simple. Fusion has occurred naturally for billions of years from the farthest reaches of the universe to our very own Milky Way. Simply put, fusion is the process that powers the stars.

The energy that comes from the process relies on the joining, or "fusing," of two atoms to form a new molecule. When this larger, relatively unstable molecule splits apart, it releases energy. The most common fusion process occurs with tritium and deuterium as its fuel. Tritium and deuterium are just big names for different isotopes of hydrogen.

So we know that tritium and deuterium fuse and then split apart to create energy. But where does the energy come from? First, let's look at the actual process of fusion. As we all now know, tritium atoms have 1 proton and 2 neutrons, and deuterium atoms have 1 proton and 1 neutron. So when a deuterium and a tritium bang together, or "fuse," a molecule with 2 protons and 3 neutrons is formed. This molecule is inherently unstable, so it splits apart. But it doesn't just revert to a tritium and deuterium. Instead, it splits into an extremely stable helium atom (2 neutrons and 2 protons). The extra neutron is simply released altogether.

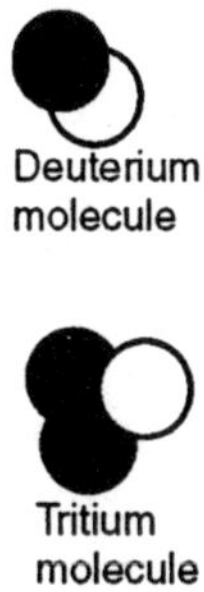


**Fig.** Isotopes of Hydrogen

There are two explanations for the production of energy, one general and one mathematical. The general explanation is that because the process results in the formation of helium, and because helium is more stable than deuterium and tritium, the system has gone from a state of high potential energy to low potential energy, and thus kinetic energy is given off.

A more mathematical explanation is derived from Albert Einstein's famous equation, $E=mc^2$. Otherwise, just accept the fact that the total mass of a helium atom and a neutron (the products of the reaction) is less than the total mass of a deuterium atom and a tritium atom (the reactants). The "missing mass" has in fact been converted to energy based on Einstein's equation.

## The Availability of Fusion Power

The future of fusion power, tritium and deuterium are virtually inexhaustible resources. There is 1 deuterium atom in every 6500 hydrogen atoms in ordinary seawater. Tritium, the other reactant, can be bred from lithium, an element common in the earth's crust.

Thus, theoretically, humans can continue to utilize fusion power indefinitely. After all, the amount of seawater on earth is practically unlimited.

Although fusion power sounds extremely attractive right now, it still has not been developed into a viable energy source. The reason is because it isn't as simple as it sounds. Tritium and deuterium atoms don't (thank goodness) randomly collide and give off energy. They must be heated up to extremely high temperatures (around 100 million degrees!) in order for fusion to take place.

So why can't we just heat them up? Well, when matter is present at the high temperatures necessary for fusion, it changes into another form. We're all familiar with the 3 main states of matter: solid, liquid, and gas. Each succeeding state is more energetic than the last. The atoms of a substance in the gaseous state move about much more than those of a substance in the solid state. Plasma, what has been called the "fourth state of matter," consists of a cloud of charged particles and is the most energetic of the four states of matter. Plasma is the most common form of matter in the universe. In fact, stars are made mostly of plasma. Examples of this unique state of matter here on earth include lightning, neon signs, and ordinary flames.

Because plasma exists only at such high temperatures, the electrons in the atoms of a substance in the plasmid state have enough potential energy to "break free" of the nucleus. Thus, the plasma is said to be made up of a cloud of positively and negatively charged particles. The positive ions are the atoms minus the electrons (and thus have a net positive charge). The negative particles are of course the electrons themselves that have been stripped from the atoms.

One characteristic of matter in the plasma state is how difficult it is to confine. Naturally, anything present at such high temperatures is very energetic and therefore moves around a great deal, and anything that moves around a lot is hard to confine. The containment problem has become one of the biggest setbacks to the use of fusion power on earth.

## Methods of Containing Plasma

There are currently three known methods of confining and heating plasma.

The first method, gravity, is the way stars throughout the galaxy keep the plasma confined long enough to produce significant energy outputs. The compression of the plasma by gravitational forces and the subsequent fusion energy produced keep the system heated and continue fusion reactions. The plasma confined by this system lasts upwards of 30 million years. However, scientists here on earth are not yet able to recreate the amazing confining forces characteristic of gravity-bound fusion systems. the extremely famous photograph of M16, a newly forming star system.

The second method, inertia, utilizes laser beams to compress the plasma. Unfortunately, this method requires extreme precision (the entire system is less than 10 cm), and the plasma produced by these means typically lasts for less than 1 tenth of a nanosecond ($10^{-7}$ seconds), an extremely brief time. A piece of equipment at the forefront of inertia-driven fusion reactions is the Glass Nova Laser, currently housed at The Lawrence Livermore National Laboratory (LLNL).

The third method utilizes magnetic fields to compress the plasma. An electromagnetic field encapsulates the plasma chamber with currents that flow in the shape of a helix. The "tokamak" is a type of magnetic field confinement chamber which today is seen as having one of the most promising prospects for harnessing fusion energy.

## Chapter 11

# Nuclear Fission

Nuclear fission is the splitting of the nucleus of an atom into parts (lighter nuclei) often producing photons (in the form of gamma rays), free neutrons and other subatomic particles as by-products. Fission of heavy elements is an exothermic reaction which can release large amounts of energy both as electromagnetic radiation and as kinetic energy of the fragments (heating the bulk material where fission takes place). Fission is a form of elemental transmutation because the resulting fragments are not the same element as the original atom.

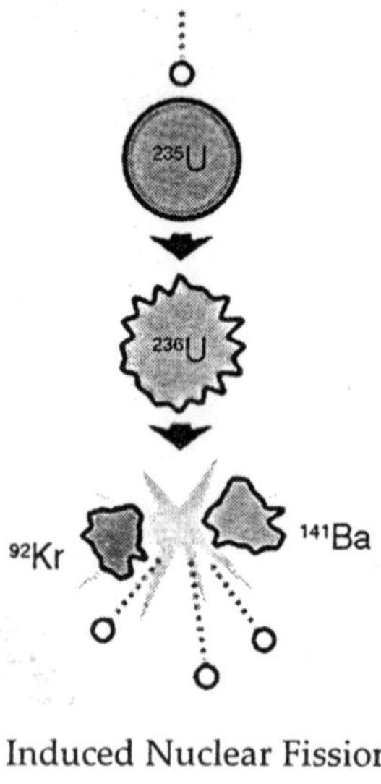


Fig. An Induced Nuclear Fission Event.

Nuclear fission produces energy for nuclear power and to drive the explosion of nuclear weapons. Both uses are made possible because certain substances called nuclear fuels undergo fission when struck by free neutrons and in turn generate neutrons when they break apart. This makes possible a self-sustaining chain reaction that releases energy at a controlled rate in a nuclear reactor or at a very rapid uncontrolled rate in a nuclear weapon. The amount of free energy contained in nuclear fuel is millions of times the amount of free energy contained in a similar mass of chemical fuel such as gasoline, making nuclear fission a very tempting source of energy; however, the products of nuclear fission are radioactive and remain so for significant amounts of time, giving rise to a nuclear waste problem. Concerns over nuclear waste accumulation and over the destructive potential of nuclear weapons may counterbalance the desirable qualities of fission as an energy source, and give rise to ongoing political debate over nuclear power.

## PHYSICAL OVERVIEW

Nuclear fission differs from other forms of radioactive decay in that it can be harnessed and controlled via a chain reaction: free neutrons released by each fission event can trigger yet more events, which in turn release more neutrons and cause more fissions. Chemical isotopes that can sustain a fission chain reaction are called nuclear fuels, and are said to be fissile. The most common nuclear fuels are $^{235}U$ (the isotope of uranium with an atomic mass of 235 and of use in nuclear reactors) and $^{239}Pu$ (the isotope of plutonium with an atomic mass of 239). These fuels break apart into a range of chemical elements with atomic masses near 100 (fission products). Most nuclear fuels undergo spontaneous fission only very slowly, decaying mainly via an alpha/beta decay chain over periods of millennia to eons. In a nuclear reactor or nuclear weapon, most fission events are induced by bombardment with another particle such as a neutron.

Typical fission events release about two hundred million eV of energy for each fission event. By contrast, most chemical oxidation reactions (such as burning coal or TNT) release at most a few eV per event, so nuclear fuel contains at least ten million times more usable energy than does chemical fuel. The energy of nuclear fission is released as kinetic energy of the fission products and fragments, and as electromagnetic radiation in the form of gamma rays; in a nuclear reactor, the energy is converted to heat as the particles and gamma rays collide with the atoms that make up the reactor and its working fluid, usually water or occasionally heavy water.

Nuclear fission of heavy elements produces energy because the specific binding energy (binding energy per mass) of intermediate-mass nuclei with atomic numbers and atomic masses close to $^{61}$Ni and $^{56}$Fe is greater than the specific binding energy of very heavy nuclei, so that energy is released when heavy nuclei are broken apart.

The total rest masses of the fission products (Mp) from a single reaction is less than the mass of the original fuel nucleus (M). The excess mass Äm = M - Mp is the invariant mass of the energy that is released as photons (gamma rays) and kinetic energy of the fission fragments, according to the mass-energy equivalence formula $E = mc^2$.

In nuclear fission events the nuclei may break into any combination of lighter nuclei, but the most common event is not fission to equal mass nuclei of about mass 120; the most common event (depending on isotope and process) is a slightly unequal fission in which one daughter nucleus has a mass of about 90 to 100 u and the other the remaining 130 to 140 u. Unequal fissions are energetically more favourable because this allows one product to be closer to the energetic minimum near mass 60 u (only a quarter of the average fissionable mass), while the other nucleus with mass 135 u is still not far out of the range of the most tightly bound nuclei (another statement of this, is that the atomic binding energy curve is slightly steeper to the left of mass 120 u than to the right of it).

The variation in specific binding energy with atomic number is due to the interplay of the two fundamental forces acting on the component nucleons (protons and neutrons) that make up the nucleus. Nuclei are bound by an attractive strong nuclear force between nucleons, which overcomes the electrostatic repulsion between protons. However, the strong nuclear force acts only over extremely short ranges, since it follows a Yukawa potential. For this reason large nuclei are less tightly bound per unit mass than small nuclei, and breaking a very large nucleus into two or more intermediate-sized nuclei releases energy.

Because of the short range of the strong binding force, large nuclei must contain proportionally more neutrons than do light elements, which are most stable with a 1-1 ratio of protons and neutrons. Extra neutrons stabilize heavy elements because they add to strong-force binding without adding to proton-proton repulsion. Fission products have, on average, about the same ratio of neutrons and protons as their parent nucleus, and are therefore usually unstable because they have proportionally too many neutrons compared to stable isotopes of similar mass. This is the fundamental cause of the problem of radioactive high level waste from nuclear reactors. Fission products tend to be beta emitters, emitting fast-moving electrons to conserve electric charge as excess neutrons convert to protons inside the nucleus of the fission product atoms.

The most common nuclear fuels, $^{235}U$ and $^{239}Pu$, are not major radiologic hazards by themselves: $^{235}U$ has a half-life of approximately 700 million years, and although $^{239}Pu$ has a half-life of only about 24,000 years, it is a pure alpha particle emitter and hence not particularly dangerous unless ingested. Once a fuel element has been used, the remaining fuel material is intimately mixed with highly radioactive fission products that emit energetic beta particles and gamma rays. Some fission products have half-lives as short as seconds; others have half-lives of tens of thousands of years, requiring long-term storage in facilities such as Yucca mountain until the fission products decay into non-radioactive stable isotopes.

## CHAIN REACTIONS

1. A uranium-235 atom absorbs a neutron and fissions into two new atoms (fission fragments), releasing three new neutrons and some binding energy.

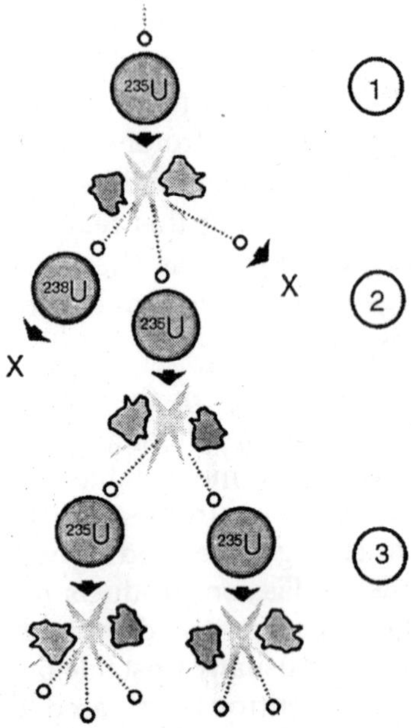


**Fig.** A schematic nuclear fission chain reaction.

2. One of those neutrons is absorbed by an atom of uranium-238 and does not continue the reaction. Another neutron is simply lost and does not collide with anything, also not continuing the reaction. However one neutron does collide with an atom of uranium-235, which then fissions and releases two neutrons and some binding energy.
3. Both of those neutrons collide with uranium-235 atoms, each of which fissions and releases between one and three neutrons, which can then continue the reaction.

Many heavy elements, such as uranium, thorium, and plutonium, undergo both spontaneous fission, a form of radioactive decay and *induced fission*, a form of nuclear reaction. Elemental isotopes that undergo induced fission when struck by a free neutron are called fissionable; isotopes that undergo fission when struck by a thermal, slow moving neutron are also called fissile. A few particularly fissile and readily obtainable isotopes (notably $^{235}U$ and $^{239}Pu$) are called nuclear fuels because they can sustain a chain reaction and can be obtained in large enough quantities to be useful.

All fissionable and fissile isotopes undergo a small amount of spontaneous fission which releases a few free neutrons into any sample of nuclear fuel. Such neutrons would escape rapidly from the fuel and become a free neutron, with a half-life of about 15 minutes before they decayed to protons and beta particles. However, neutrons almost invariably impact and are absorbed by other nuclei in the vicinity long before this happens (newly-created fission neutrons are moving at about 7percent of the speed of light, and even moderated neutrons are moving at about 8 times the speed of sound). Some neutrons will impact fuel nuclei and induce further fissions, releasing yet more neutrons. If enough nuclear fuel is assembled into one place, or if the escaping neutrons are sufficiently contained, then these freshly generated neutrons outnumber the neutrons that escape from the assembly, and a *sustained nuclear chain reaction* will take place.

An assembly that supports a sustained nuclear chain reaction is called a critical assembly or, if the assembly is almost entirely made of a nuclear fuel, a critical mass. The word "critical" refers to a cusp in the behaviour of the differential equation that governs the number of free neutrons present in the fuel: if less than a critical mass is present, then the amount of neutrons is determined by radioactive decay, but if a critical mass or more is present, then the amount of neutrons is controlled instead by the physics of the chain reaction. The actual mass of a *critical mass* of nuclear fuel depends strongly on the geometry and surrounding materials.

Not all fissionable isotopes can sustain a chain reaction. $^{238}U$, the most abundant form of uranium, is fissionable but not fissile: it undergoes induced fission when impacted by an energetic neutron with over 1 MeV of kinetic energy. But too few of the neutrons produced by $^{238}U$ fission are energetic enough to induce further fissions in $^{238}U$, so no chain reaction is possible with this isotope. Instead, bombarding $^{238}U$ with slow neutrons causes it to absorb them (becoming $^{239}U$) and decay by beta emission to $^{239}Np$ which then decays again by the same process to $^{239}Pu$; that process is used to manufacture $^{239}Pu$ in breeder reactors, but does not contribute to a neutron chain reaction.

Fissionable, non-fissile isotopes can be used as fission energy source even without a chain reaction. Bombarding $^{238}U$ with fast neutrons induces fissions, releasing energy as long as the external neutron source is present. That effect is used to augment the energy released by modern thermonuclear weapons, by jacketing the weapon with $^{238}U$ to react with neutrons released by nuclear fusion at the centre of the device.

## FISSION REACTORS

Critical fission reactors are the most common type of nuclear reactor. In a critical fission reactor, neutrons produced by fission of fuel atoms are used to induce yet more fissions, to sustain a controllable amount of energy release. Devices that produce engineered but non-self-sustaining fission reactions are *subcritical fission reactors*. Such devices use radioactive decay or particle accelerators to trigger fissions.

Critical fission reactors are built for three primary purposes, which typically involve different engineering trade-offs to take advantage of either the heat or the neutrons produced by the fission chain reaction:

- *Power reactors* are intended to produce heat for nuclear power, either as part of a generating station or a local power system such as a nuclear submarine.
- *Research reactors* are intended to produce neutrons and/or activate radioactive sources for scientific, medical, engineering, or other research purposes.
- *Breeder reactors* are intended to produce nuclear fuels in bulk from more abundant isotopes. The better known fast breeder reactor makes $^{239}Pu$ (a nuclear fuel) from the naturally very abundant $^{238}U$ (not a

nuclear fuel). Thermal breeder reactors previously tested using $^{232}Th$ continue to be studied and developed.

While, in principle, all fission reactors can act in all three capacities, in practice the tasks lead to conflicting engineering goals and most reactors have been built with only one of the above tasks in mind. Power reactors generally convert the kinetic energy of fission products into heat, which is used to heat a working fluid and drive a heat engine that generates mechanical or electrical power. The working fluid is usually water with a steam turbine, but some designs use other materials such as gaseous helium. Research reactors produce neutrons that are used in various ways, with the heat of fission being treated as an unavoidable waste product. Breeder reactors are a specialized form of research reactor, with the caveat that the sample being irradiated is usually the fuel itself, a mixture of $^{238}U$ and $^{235}U$.

For a more detailed description of the physics and operating principles of critical fission reactors. For a description of their social, political, and environmental aspects.

## FISSION BOMBS

One class of nuclear weapon, a *fission bomb* (not to be confused with the *fusion bomb*), otherwise known as an *atomic bomb* or *atom bomb*, is a fission reactor designed to liberate as much energy as possible as rapidly as possible, before the released energy causes the reactor to explode (and the chain reaction to stop). Development of nuclear weapons was the motivation behind early research into nuclear fission: the Manhattan Project of the U.S. military during World War II carried out most of the early scientific work on fission chain reactions, culminating in the Little Boy and Fat Man and Trinity bombs that were exploded over test sites, the cities Hiroshima, and Nagasaki, Japan in August of 1945.

Even the first fission bombs were thousands of times more explosive than a comparable mass of chemical explosive. Little Boy weighed a total of about four tons (of which 60 kg was nuclear fuel) and was 11 feet long; it also yielded an explosion equivalent to about 15,000 tons of TNT, destroying a large part of the city of Hiroshima. Modern nuclear weapons (which include a thermonuclear *fusion* as well as one or more fission stages) are literally hundreds of times more energetic for their weight than the first pure fission atomic bombs, so that a modern single missile warhead bomb weighing less than 1/8th as much as Little Boy has a yield of 475,000 tons of TNT, and could bring destruction to 10 times the city area.

While the fundamental physics of the fission chain reaction in a nuclear weapon is similar to the physics of a controlled nuclear reactor, the two types of device must be engineered quite differently. It is impossible to convert a nuclear reactor to cause a true nuclear explosion, or for a nuclear reactor to explode the way a nuclear explosive does, (though partial fuel meltdowns and steam explosions have occurred), and similarly difficult to extract useful power from a nuclear explosive (though at least one rocket propulsion system, Project Orion, was intended to work by exploding fission bombs behind a massively padded vehicle!).

The strategic importance of nuclear weapons is a major reason why the technology of nuclear fission is politically sensitive. Viable fission bomb designs are, arguably, within the capabilities of bright undergraduates ( being incredibly simple, but nuclear fuel to realise the designs is thought to be difficult to obtain being rare.

While many believe that Ernest Rutherford became the first person to deliberately split the atom by bombarding nitrogen with naturally occurring alpha particles from radioactive material and observing a proton emitted with energy higher than the alpha particle, this is incorrect. Rutherford forced protons out of the nucleus but did not split the nucleus, as a fission reaction requires. His students1932 John Cockcroft and Ernest Walton, working under Rutherford's direction, attempted to split the nucleus by entirely artificial means, using a particle accelerator to bombard lithium with protons thereby producing two alpha particles. This still did not result in nuclear fission.

Results of the bombardment of uranium by neutrons had proved interesting and puzzling. First studied by Enrico Fermi and his colleagues in 1934, they were not properly interpreted until several years later.

After the Fermi publication, Lise Meitner, Otto Hahn and Fritz Strassmann began performing similar experiments in Germany. Meitner, an Austrian Jew, lost her citizenship with the Anschluss in 1938. She fled and wound up in Sweden, but continued to collaborate by mail and through meetings with Hahn in Sweden. By coincidence her nephew Otto Robert Frisch, also a refugee, was also in Sweden when Meitner received a letter from Hahn describing his chemical proof that some of the product of the bombardment of uranium with neutrons was barium (barium's atomic weight is half that of uranium). Frisch was skeptical, but Meitner believed Hahn was too good a chemist to have made a mistake. According to Frisch:

> No larger fragments than protons or helium nuclei (alpha particles) had ever been chipped away from nuclei, and to chip off a large number not nearly enough energy was available. Nor was it possible that the uranium nucleus could have been cleaved right across. A nucleus was not like a brittle solid that can be cleaved or broken; George Gamow had suggested early on, and Bohr had given good arguments that a nucleus was much more like a liquid drop. Perhaps a drop could divide itself into two smaller drops in a more gradual manner, by first becoming elongated, then constricted, and finally being torn rather than broken in two? We knew that there were strong forces that would resist such a process, just as the surface tension of an ordinary liquid drop tends to resist its division into two smaller ones. But nuclei differed from ordinary drops in one important way: they were electrically charged, and that was known to counteract the surface tension.
>
> The charge of a uranium nucleus, we found, was indeed large enough to overcome the effect of the surface tension

almost completely; so the uranium nucleus might indeed resemble a very wobble unstable drop, ready to divide itself at the slightest provocation, such as the impact of a single neutron. But there was another problem. After separation, the two drops would be driven apart by their mutual electric repulsion and would acquire high speed and hence a very large energy, about 200 MeV in all; where could that energy come from?...Lise Meitner... worked out that the two nuclei formed by the division of a uranium nucleus together would be lighter than the original uranium nucleus by about one-fifth the mass of a proton. Now whenever mass disappears energy is created, according to Einstein's formula $E=mc^2$, and one-fifth of a proton mass was just equivalent to 200MeV. So here was the source for that energy; it all fitted!

The basic discovery and chemical proof of Otto Hahn and Fritz Strassmann that an isotope of barium was produced by neutron bombardment of uranium was published in a paper in Germany in the Journal *Naturwissenschaften,* January 6, 1939) and earned Hahn a Nobel Prize.

Frisch rapidly confirmed experimentally by means of a cloud chamber that the uranium atom had indeed been split by the action of neutrons. A fundamental idea of this experiment was suggested to Frisch by George Placzek. Two papers were mailed to England on January 16, 1939, the first on the interpretation of the barium appearance as atom splitting by Meitner and Frisch, the second on the experimental confirmation by Frisch (strangely omitting Placzek's important contribution, however). The first paper appeared on February 11, the second on February 28.

Meitner and Frisch's theory and mathematical proof of Hahn's discovery and chemical proof of barium products from the bombardment of uranium was the foundation of the later research on nuclear fission. The awarding of the 1944 Nobel Prize in Chemistry to Hahn alone is a longstanding controversy.

On January 16, 1939, Niels Bohr of Copenhagen, Denmark, arrived in the United States to spend several months in

Princeton, New Jersey, and was particularly anxious to discuss some abstract problems with Albert Einstein. Just before Bohr left Denmark, Frisch and Meitner gave him their calculations.

Bohr had promised to keep the Meitner/Frisch paper secret until it was published to preserve priority, but on the boat he discussed it with Léon Rosenfeld, and forgot to tell him to keep it secret. Rosenfeld immediately upon arrival told everyone at Princeton University, and from them the news spread by word of mouth to neighbouring physicists including Enrico Fermi at Columbia University. Fermi upon traveling to receive the Nobel Prize for his earlier work. headed to the USA rather than return to Fascist Italy with his Jewish wife. As a result of conversations among Fermi, John R. Dunning, and G. B. Pegram, a search was undertaken at Columbia for the heavy pulses of ionization that would be expected from the flying fragments of the uranium nucleus. On January 26, 1939, there was a conference on theoretical physics at Washington, D.C., sponsored jointly by the George Washington University and the Carnegie Institution of Washington. Before the meeting in Washington was over, several other experiments to confirm fission had been initiated, and positive experimental confirmation was reported.

Frédéric Joliot-Curie's team in Paris discovered that secondary neutrons are released during uranium fission thus making a chain reaction feasible. About two neutrons being emitted with nuclear fission of uranium was verified independently by Leo Szilard and Walter Zinn. The number of neutrons emitted with nuclear fission of 235uranium was then reported at 3.5/fission, and later corrected to 2.6/fission by Frédéric Joliot-Curie, Hans von Halban and Lew Kowarski.

"Chain reactions" at that time were a known phenomenon in *chemistry*, but the analogous process in nuclear physics using neutrons had been foreseen as early as 1933 by Leo Szilard, although Szilard at that time had no idea with what materials the process might be initiated. Szilard, a Hungarian born Jew, also fled mainland Europe after Hitler's rise, eventually landing in the US.

In the summer Fermi and Szilard proposed the idea of a nuclear reactor (pile) with natural uranium as fuel and graphite as moderator of neufron energy.

In August Hungarian-Jewish refugees Szilard, Teller and Wigner persuaded Austrian-Jewish refugee Einstein to warn President Roosevelt of the German menace. The letter suggested the possibility of uranium bomb deliverable by ship. The President received it on 1939.10.11 shortly after WWII began.

In England James Chadwick proposed an atomic bomb utilizing natural uranium based on a paper by Rudolf Peierls with the mass needed for critical state being 30-40 tons.

In December, Heisenberg delivered a report to the Germany Department of War on the possibility of a uranium bomb.

In Birmingham, England Otto Robert Frisch teamed up with Rudolf Peierls who had also fled German anti-Jewish race laws. They conceived the idea of utilizing a purified isotope of uranium, uranium-235, and worked out that an enriched uranium bomb could have a critical mass of only 600 g. instead of tons, and that the resulting explosion would be tremendous. (the amount actually turned out to be 15 kg.) In February 1940 they delivered the Frisch-Peierls memorandum, however, they were officially considered "enemy aliens" at the time.

Glenn Seaborg, Joe Kennedy, Art Wahl and Italian-Jewish refugee Emilio Segre discovered plutonium and determined it to be fissionable like U-235. (Lawrence controversially dropped Segre's pay by half when he learned he was trapped in the US by Mussolini's race laws.)

On June 28 1941, the Office of Scientific Research and Development was formed to mobilize scientific resources and apply the results of research to national defence. In September Fermi assembled his first nuclear pile in an attempt to create a slow neutron induced chain reaction in uranium. but the experiment failed.

Producing a fission chain reaction in uranium fuel is far from trivial. Early nuclear reactors did not use isotopically enriched uranium, and in consequence they were required to use large quantities of highly purified graphite as neutron moderation materials. Use of ordinary water (as opposed to heavy water) in nuclear reactors requires enriched fuel— the partial separation and relative enrichment of the rare $^{235}U$ isotope from the far more common $^{238}U$ isotope. Typically, reactors also require inclusion of extremely chemically pure neutron moderator materials such as deuterium (in heavy water), helium, beryllium, or carbon, usually as the graphite. (The high purity is required because many chemical impurities such as the boron-10 component of natural boron, are very strong neutron absorbers and thus poison the chain reaction.)

Production of such materials at industrial scale had to be solved for nuclear power generation and weapons production to be accomplished. Up to 1940, the total amount of uranium metal produced in the USA was not more than a few grams and even this was of doubtful purity; of metallic beryllium not more than a few kilograms; concentrated deuterium oxide (heavy water) not more than a few kilograms; and finally carbon had never been produced in quantity with anything like the purity required of a moderator.

The problem of producing large amounts of high purity uranium was solved by Frank Spedding using the thermite process. Ames Laboratory was established in 1942 to produce the large amounts of natural (unenriched) uranium that would be necessary for the research to come. The success of the Chicago Pile-1 which used unenriched (natural) uranium, like all of the atomic "piles" which produced the plutonium for the atomic bomb, was also due specifically to Szilard's realization that very pure graphite could be used for the moderator of even natural uranium "piles". In wartime Germany, failure to appreciate the qualities of very pure graphite led to reactor designs dependent on heavy water, which in turn was denied the Germans by Allied attacks in

Norway, where heavy water was produced. These difficulties prevented the Nazis from building a nuclear reactor capable of criticality during the war.

Unknown until 1972, when French physicist Francis Perrin discovered the Oklo Fossil Reactors, nature had beaten humans to the punch by engaging in large-scale uranium fission chain reactions, some 2,000 million years in the past. This ancient process was able to use normal water as a moderator, only because 2,000 million years in the past, natural uranium was "enriched" with the shorter-lived fissile isotope $^{235}U$, as compared with the natural uranium available today.

## Natural Nuclear Fission Reactor

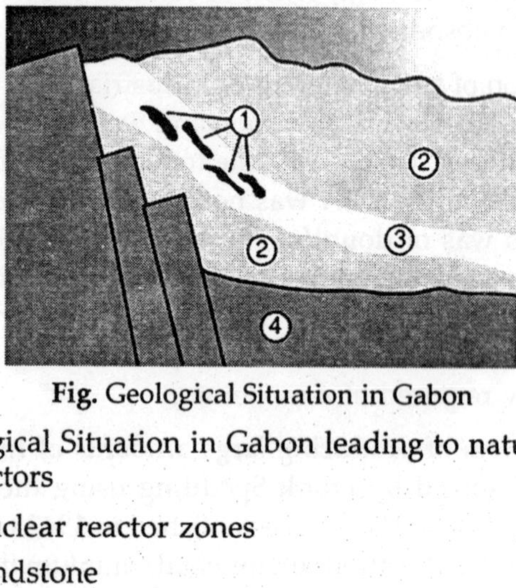


**Fig.** Geological Situation in Gabon

Geological Situation in Gabon leading to natural nuclear fission reactors

1. Nuclear reactor zones
2. Sandstone
3. Uranium ore layer
4. Granite

A natural nuclear fission reactor is a uranium deposit where analysis of isotope ratios has shown that self-sustaining nuclear chain reactions have occurred. The existence of this phenomenon was discovered in 1972 by French physicist Francis Perrin. The conditions under which a natural nuclear reactor could exist were predicted at the University of Arkansas by Paul Kuroda in 1956. The conditions found at Oklo were very similar to what Kuroda predicted.

At the only known location, three ore deposits at Oklo in Gabon, sixteen sites have been discovered so far at which self-sustaining nuclear fission reactions took place approximately 1.5 billion years ago, and ran for a few hundred thousand years, averaging 100 kW of power output during that time.

In May 1972 at the Pierrelatte uranium enrichment facility in France, routine mass spectrometry comparing $UF_6$ samples from the Oklo Mine, located in Gabon, Central Africa, showed a discrepancy in the amount of the $^{235}U$ isotope. Normally the concentration is 0.7202percent; these samples had only 0.7171percent – a significant difference. This discrepancy required explanation, as all uranium handling facilities must meticulously account for all fissionable isotopes to assure that none are diverted for weapons purposes. Thus the French Commissariat à l'énergie atomique began an investigation. A series of measurements of the relative abundances of the two most significant isotopes of the uranium mined at Oklo showed anomalous results compared to those obtained for uranium from other mines. Further investigations into this uranium deposit discovered uranium ore with a $^{235}U$ to $^{238}U$ ratio as low as 0.440percent.

## Fission Product Isotope Signatures

### *Nd*

Neodymium and other elements were found with isotopic compositions different from what is customarily found on Earth. Natural neodymium contains 27percent $^{142}Nd$; the Nd at Oklo contained less than 6percent but contained more $^{143}Nd$.

Subtracting the natural isotopic Nd abundance from the Oklo-Nd, the isotopic composition matched that produced by the fissioning of $^{235}U$.

A diagram showing the isotope signatures of natural neodymium and fission product neodymium from U-235 which had been subjected to thermal neutrons. Note that the

Ce-142 (a long lived beta emitter) has not had time to decay to Nd-142 over the time since the reactors stopped working.

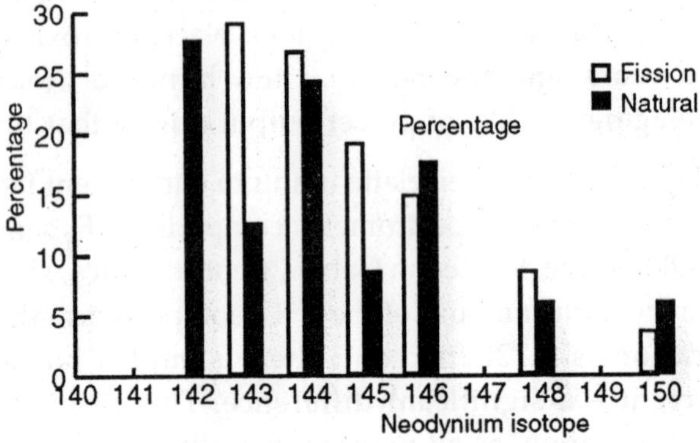


***Ru***

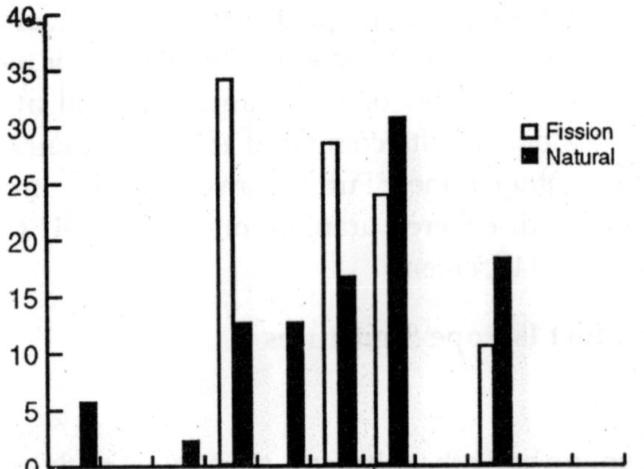


Similar investigations into the isotopic ratios of ruthenium at Oklo found a much higher $^{99}$Ru concentration than expected (27-30percent vs. 12.7percent). This anomaly could be explained by the decay of $^{99}$Tc to $^{99}$Ru. In the bar chart below the normal natural isotope signature of ruthenium is compared with that for fission product ruthenium which is the result of the fission of $^{235}$U with thermal neutrons. It is clear that the fission ruthenium has a different isotope signature. The level of $^{100}Ru$ in the fission product mixture is low because of a long lived (half life = $10^{19}$ years) isotope of molybdenum. On the time scale of when the reactors were in operation very little decay to $^{100}Ru$ will have occurred.

A diagram showing the isotope signatures of natural ruthenium and fission product ruthenium from U-235 which had been subjected to thermal neutrons. Note that the Mo-100 (a long lived double beta emitter) has not had time to decay to Ru-100 over the time since the reactors stopped working.

Other observations led to the same conclusion, and on September 25, 1972, the CEA announced their finding that self-sustaining nuclear chain reactions had occurred on Earth about 2 billion years ago. Later, other natural nuclear fission reactors were discovered in the region.

## Mechanism of The Reactors

The natural nuclear reactor formed when a uranium-rich mineral deposit became inundated with groundwater that acted as a neutron moderator, and a strong chain reaction took place. The water moderator would boil away as the reaction increased, slowing it back down again and preventing a meltdown. The fission reaction was sustained for hundreds of thousands of years. Fission of uranium normally produces five known isotopes of the fission-product gas xenon; all five have been found trapped within novel aluminium foams in the remnants of the natural reactor, in varying concentrations. This points to on-again, off-again reactor operation. The specific concentrations of xenon isotopes, found trapped in mineral formations 2 billion years later, make it possible to calculate the specific time intervals of reactor operation: approximately 2 hours and 30 minutes.

It is estimated that secondary enrichment of the uranium in centimeter- to meter-sized veins consumed about six tons of $^{235}U$ and elevated temperatures to a few hundred degrees Celsius. Remarkably, the non-volatile fission products have only moved a few centimeters in the veins during the last 1.5 billion years. This offers a case study of how radioactive isotopes migrate through the earth's crust—a significant area of controversy as opponents of geologic nuclear waste disposal fear that releases from stored waste could end up in water supplies or be carried into the environment.

A key factor that made the reaction possible was that, at the time the reactor went critical, the fissile isotope $^{235}U$ made up about 3percent of the natural uranium, which is comparable to the amount used in some of today's reactors. (The remaining 97percent was non-fissile $^{238}U$.) Because $^{235}U$ has a shorter half life than $^{238}U$, and thus decays more rapidly, the current abundance of $^{235}U$ in natural uranium is about 0.7percent. A natural nuclear reactor is therefore no longer possible on Earth.

Another factor which probably contributed to the operation of the Oklo natural nuclear reactor was the fact that uranium is soluble in water only in the presence of oxygen. Increasing oxygen content in the earth's atmosphere helps explain why this natural reactor started operation sometime around 2 billion years ago and not before, since the percentage abundance of fissionable $^{235}U$ was at least 3percent or higher at all times before reactor startup. Rising oxygen levels during the aging of earth may have allowed a high enough concentration of uranium to accumulate to enhance the possibility of reactor startup.

When oxygenated water trickled over uranium deposits and picked up a few ppm of uranium, uranium may have been re-deposited and concentrated due to the action of algae and geological placer formations more than once. Other geological action, such as uplifting and fracturing of the concentrated uranium ore formations, may have caused further concentrations into a compacted vein. Without the new aerobic environment available on earth at the time, these concentrations probably couldn't have taken place.

The natural reactor of Oklo can also be used to check if the fine-structure constant á might have changed over time. Alex Shlyakhter proposed in 1976 to measure the abundance of $^{149}Sm$ to estimate the cross section for neutron capture of this isotope at the time of the reactions and check it against the present value. Measurements show that the constant has not changed

## Relation to Yucca Mountain and other Geologic Repositories

The deep geological repository concept involves the encapsulation of used fuel in long-lived engineered containers which are then placed and sealed within excavated rooms in a naturally occurring geological formation at a design depth of 500 to 1000 metres below ground surface.

The ability of natural geologic barriers to isolate radioactive waste is demonstrated by the Oklo reactors. During their long reaction period about 5.4 tonnes of fission products as well as 1.5 tonnes of plutonium together with other transuranic elements were generated in the orebody. This plutonium and the other transuranics remained immobile until the present day. This is quite remarkable because ground water had ready access to the deposits and they were not in a chemically inert form, such as glass.

Thus the only known example of underground nuclear waste disposal was successful over a long period in spite of the characteristics of the site. Such a water-logged, sandstone/shale structure would not be considered for disposal of modern toxic wastes, nuclear or otherwise, although the clays and bitumen present played an important part in containing the material.

## BASIC NUCLEAR FISSION

Nuclear fission is the process of splitting atoms, or fissioning them. This topic will explain the basics of nuclear fission. Everyone's played with marbles at one time or another, right? Well, imagine about 200 marbles lying on a flat surface, all jumbled together, and roughly forming a circle. What would happen if someone took another marble and threw it at them? They would fly all around in different directions and groups, right? That is exactly what happens in nuclear fission. The filled circle is like an atom's nucleus. The marble being thrown is like a "neutron bullet". The only differences are that the marbles are protons and neutrons and the protons and neutrons aren't in a filled circle, but in the actual atom are in the shape of a sphere. Of course, an atom is also a bit more complicated than a pack of marbles.

## Choosing the Bullet

When we spoke about the marble analogy earlier, we said that the marble being thrown is a like a "neutron bullet". But what does this mean, and why not use another type of particle to "throw" at a nucleus to fission it? First, what particles with distinct mass are available to launch at a nucleus? Think back to our lesson on radioactivity. Recall that two particles emitted by radioactive elements are the $\alpha$ particle and the neutron. (There are other particles emitted too, but they are generally much smaller than the neutron and the $\alpha$ particle.) Recall that the $\alpha$ particle is essentially a $^4$He nucleus. Now, let's review the structure of an atom. Remember that an atomic nucleus is made up of positive protons and neutral neutrons? Because of this, the nucleus carries an overall positive charge. So, if we were to launch another particle with a positive charge at a nucleus, it wouldn't get there. Why wouldn't it get there? The answer lies in magnetism. Have you ever used magnets? If you have, you'd know that two like poles of a magnet repel each other. A positive particle and the positive nucleus would repel each other in the same way. The $\alpha$ particle is positive. Why? Well, it's composed of two protons and two neutrons. Its positive protons give it a positive charge. Because it's positive, it would get repelled away from another positive nucleus. So, the only thing left is the neutron. The neutron is electrically neutral and thus would not get repelled from a positive nucleus.

## Fissile Isotopes

Fissile isotopes are isotopes of an element that can be split through fission. Only certain isotopes of certain elements are fissile. One isotope of uranium, $_{235}$U, is fissile, while another isotope, $_{238}$U, is not. Other examples of fissile elements are

$_{239}Pu$ and $_{232}Th$. An important factor affecting whether or not an atom will fission is the speed at which the bombarding neutron is moving. If the neutron is highly energetic (and thus moving very quickly), it can cause fission in some elements that a slower neutron would not. Thorium 232 requires a very fast neutron to induce fission. However, uranium 235 needs slower neutrons. If a neutron is too fast, it will pass right through a $_{235}U$ atom without affecting it at all.

## Splitting the Uranium Atom

Uranium is the principle element used in nuclear reactors and in certain types of atomic bombs. The specific isotope used is $_{235}U$. When a stray neutron strikes a $_{235}U$ nucleus, it is at first absorbed into it. This creates $_{236}U$. $_{236}U$ is unstable and this causes the atom to fission. The fissioning of $_{236}U$ can produce over twenty different products. However, the products' masses always add up to 236. The following two equations are examples of the different products that can be produced when $_{235}U$ fissions:

$$_{235}U + 1 \text{ neutron} \rightarrow 2 \text{ neutrons} + {}_{92}Kr + {}_{142}Ba + \text{ENERGY}$$

$$_{235}U + 1 \text{ neutron} \rightarrow 2 \text{ neutrons} + {}_{92}Sr + {}_{140}Xe + \text{ENERGY}$$

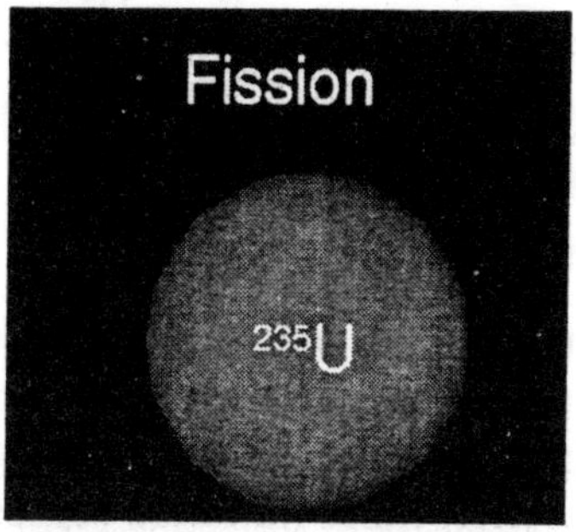


**Fig.** Fission of Uranium Atom

Let's discuss those reactions. In each of the reactions, 1 neutron splits the atom. When the atom is split, 1 additional neutron is released. This is how a chain reaction works. If more $_{235}U$ is present, those 2 neutrons can cause 2 more atoms to split. Each of those atoms releases 1 more neutron bringing the total neutrons to 4. Those **4 neutrons** can strike 4 more $_{235}U$

atoms, releasing even more neutrons. The chain reaction will continue until all the $_{235}U$ fuel is spent. This is roughly what happens in an atomic bomb. It is called a runaway nuclear reaction.

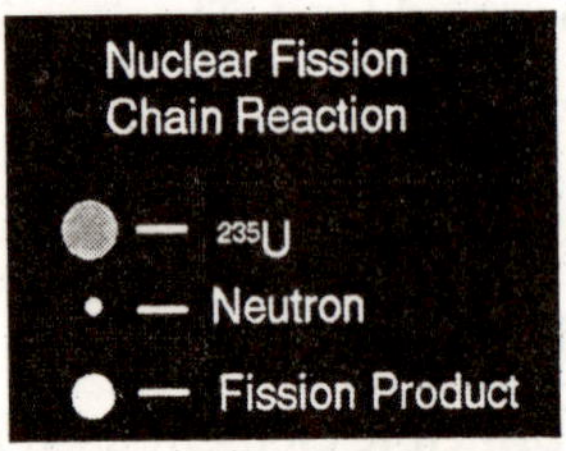


**Fig.** Fission of Uranium Atom

The fissioning of each 235U atom (red) releases more neutrons (green) that go on to fission more $_{235}U$ atoms, thus producing a chain reaction.

### *Where Does the Energy Come From?*

In the section above we described what happens when an $_{235}U$ atom fissions. We gave the following equation as an example:

$$_{235}U + 1 \text{ neutron} \rightarrow 2 \text{ neutrons} + {}_{92}Kr + {}_{142}Ba + \text{ENERGY}$$

It might have been a amazing question, "Where does the energy come from?". The mass seems to be the same on both sides of the reaction:

$$235 + 1 = 2 + 92 + 142 = 236$$

Thus, it seems that no mass is converted into energy. However, this is not entirely correct. The mass of an atom is more than the sum of the individual masses of its protons and neutrons, which is what those numbers represent. Extra mass is a result of the binding energy that holds the protons and neutrons of the nucleus together. Thus, when the uranium atom is split, some of the energy that held it together is released as radiation in the form of heat. Because energy and mass are one and the same, the energy released is also mass released. Therefore, the total mass does decrease a tiny bit during the reaction.

## THE NUCLEAR FISSION POWER PLANT

Currently, about half of all nuclear power plants are located in the US. There are many different kinds of nuclear power plants, and we will discuss a few important designs in this text. A nuclear power plant harnesses the energy inside atoms themselves and converts this to electricity. This electricity is used by all of us. By now, you should have an idea of the fission process and how it works. A nuclear power plant uses controlled nuclear fission. We will explore how a nuclear power plant operates and the manner in which nuclear reactions are controlled.

### Uranium Preparation

Earlier we talked about nuclear fission with $_{235}U$. In reality, this will not be the only isotope of uranium present in a nuclear reactor. In naturally occurring uranium deposits, less than one percent of the uranium is $_{235}U$. The majority of the uranium is $_{238}U$. $_{238}U$ is not a fissile isotope of uranium. When $_{238}U$ is struck by a loose neutron, it absorbs the neutron into its nucleus and does not fission. Thus, by absorbing loose neutrons, $_{238}U$ can prevent a nuclear chain reaction from occurring. This would be a bad thing because if a chain reaction doesn't occur, the nuclear reactions can't sustain themselves, the reactor shuts down, and millions of people are without electrical power. In order for a chain reaction to occur, the pure uranium ore must be refined to raise the concentration of $_{235}U$. This is called enrichment and is primarily accomplished through a technique called gaseous diffusion. In this process, the uranium ore is combined with fluorine to create a chemical compound called uranium hexafluoride. The uranium hexafluoride is heated and vapourizes. The heated gas is then pushed through a series of filters. Because some of the uranium hexafluoride contains 238U and some contains $_{235}U$, there is a slight difference in the weights of the individual molecules. The molecules of uranium hexafluoride containing $_{235}U$ are slightly lighter and thus pass more easily through the filters. This creates a quantity of uranium hexafluoride with a higher proportion of $_{235}U$. This is collected, the uranium is stripped from it, and the result is an enriched supply of fuel. Usually, nuclear power plants use uranium fuel that is about 4percent $_{235}U$.

### *Parts of a Nuclear Reactor - Pressurized Water Reactor (PWR):*

**Fig.** Fuel Assembly Containing a Number of Fuel Rods

A typical nuclear reactor has a few main parts. Inside the "core" where the nuclear reactions take place are the fuel rods and assemblies, the control rods, the moderator, and the coolant. Outside the core are the turbines, the heat exchanger, and part of the cooling system.

The fuel assemblies are collections of fuel rods. These rods are each about 3.5 meters (11.48 feet) long. They are each about a centimeter in diameter. These are grouped into large bundles of a couple hundred rods called fuel assemblies, which are then placed in the reactor core. Inside each fuel rod are hundreds of pellets of uranium fuel stacked end to end.

Also in the core are control rods. These rods have pellets inside that are made of very efficient neutron capturers. An example of such a material is cadmium. These control rods are connected to machines that can raise or lower them in the core. When they are fully lowered into the core, fission can not occur because they absorb free neutrons. However, when they are pulled out of the reactor, fission can start again anytime a stray neutron strikes a $_{235}$U atom, thus releasing more neutrons, and starting a chain reaction.

Another component of the reactor is the moderator. The moderator serves to slow down the high speed neutrons "flying" all around the reactor core. If a neutron is moving too fast, and thus is at a high-energy state, it passes right through the $_{235}$U nucleus. It must be slowed down to be captured by the nucleus and to induce fission. The most common moderator is water, but sometimes it can be another material.

The job of the coolant is to absorb the heat from the reaction. The most common coolant used in nuclear power plants today is water. In actuality, in many reactor designs the coolant and the moderator are one and the same. The coolant water is heated by the nuclear reactions going on inside the core. However, this heated water does not boil because it is kept at an extremely intense pressure, thus raising its boiling point above the normal 100° Celsius.

The heated water rises up and passes through another part of the reactor, the heat exchanger. The moderator/coolant water is radioactive, so it can not leave the inner reactor containment. Its heat must be transferred to non-radioactive water, which can then be sent out of the reactor shielding. This is done through the heat exchanger, which works by moving the radioactive water through a series of pipes that are wrapped around other pipes. The metallic pipes conduct the heat from the moderator to the normal water. Then, the normal water (now in steam form and intensely hot) moves to the turbine, where electricity is produced.

After the hot water has passed through the turbine, some of its energy is changed into electricity. However, the water is still very hot. It must be cooled somehow. Many nuclear power plants used steam towers to cool this water with air. These are generally the buildings that people associate with nuclear power plants. At reactors that do not have towers, the clean water is purified and dumped into the nearest body of water, and cool water is pumped in to replace it.

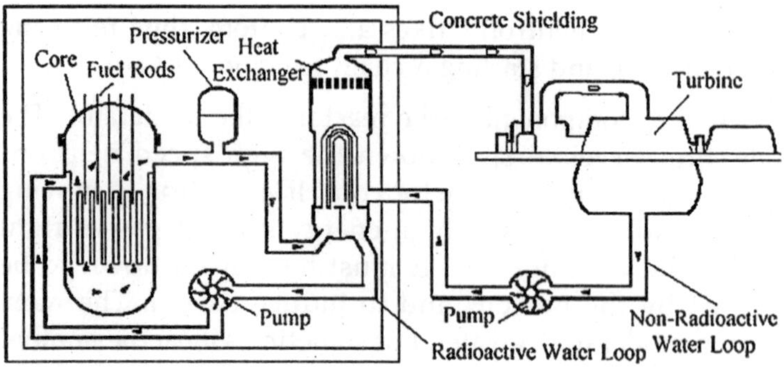


Fig. PWR Power Plant Schematic

### *From Fission to Electricity*

A nuclear power plant produces electricity in almost exactly the same way that a conventional (fossil fuel) power plant does. A conventional power plant burns fuel to create heat. The fuel is generally coal, but oil is also sometimes used. The heat is used to raise the temperature of water, thus causing it to boil. The high temperature and intense pressure steam that results from the boiling of the water turns a turbine, which then generates electricity. A nuclear power plant works the same way, except that the heat used to boil the water is produced by a nuclear fission reaction using $_{235}$U as fuel, not the combustion of fossil fuels. A nuclear power plant uses much less fuel than a comparable fossil fuel plant. A rough estimate is that it takes 17,000 kilograms of coal to produce the same amount of electricity as 1 kilogram of nuclear uranium fuel.

### *Other Types of Reactors*

Although the most common type of reactor is the Pressurized Water Reactor (PWR), many other types of reactors are also used. In the PWR, as we described earlier, there are two main water cycles. One is the water inside the core that is highly radioactive. This water's heat is transferred to other, non-radioactive water inside the second loop. This water is then used to turn a turbine. The second most popular reactor type is the Boiling Water Reactor (BRW). This type of reactor differs from the PWR in that there is only one water cycle. Radioactive water is used to turn the turbine. The major disadvantage of this is that the radioactive nuclides in the water that cause its radioactivity can be transferred to the turbine, thus causing it to become radioactive too. This produces more hazardous material that needs to be disposed of when a reactor is dismantled. However, the BWR also has a few advantages. Its core can be kept at a lower pressure.

Another type of reactor is the Heavy Water Reactor (HWR). A HWR uses heavy water as a moderator instead of normal water. Heavy water is water with deuterium, which is an isotope of hydrogen with 1 neutron. Deuterium is heavier than normal hydrogen, which has no neutrons. HWR's come in two types, pressurized and boiling, just like normal "light water" reactors. The advantage of a HWR is that un-enriched uranium fuel can be used. This is because the heavy water is a much more efficient moderator than light water. Thus, more stray neutrons can be slowed down enough to cause fission in 235U. This more efficient moderator makes up for the greater abundance of the neutron-capturing $_{238}U$.

## NUCLEAR CHAIN REACTION

In a fission reaction, besides the two fission products, a few (in the case of U-235 2.47 on the average) neutrons are emitted. The principle of chain reaction is very easy therefore: a U235 nucleus is hit by a neutron, upon which fission occurs at a high probability. The neutrons produced in this process can cause more fissions, moreover, the already mentioned 200 MeV energy is released each time. Consequently, the process becomes self-sustaining and no external neutron source is needed any longer.

If the neutrons produced from a fission induce only one new fission (because of escaping from the system or being captured inside) we say that the chain reaction is of constant intensity. If the neutrons generated from a fission reaction induce more than one fission than more and more fission reactions will take place in a unit time and the chain reaction will be diverging.

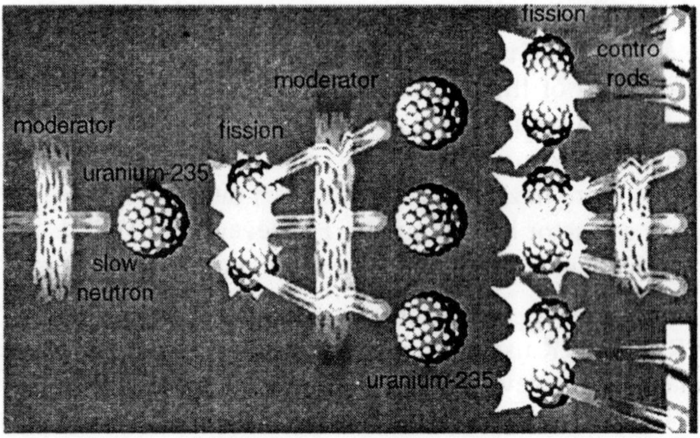


**Fig.** Nuclear chain reaction

The probability that a high energy neutron emitted in a fission reaction can directly cause fission is quite low. In order to build a system in which fast neutrons sustain the chain reaction, very highly enriched uranium is needed, which is very expensive. A more easily feasible way is to use materials that slow the neutrons down to such low energies at which the probability of causing a fission is significantly higher. These neutron slowing down materials are the so called moderators.With the aid of some adequate moderator one might as well achieve chain reaction using natural (0.7percent U235 content) uranium.

If we simplify things a little, there are two major requirements that a moderator material should meet: it should have as low atomic mass number as possible and its neutron absorbing ability should be as low as possible. In practice four elements meet these requirements: water ($H_2O$), heavy-water ($D_2O$), graphite (C) and beryllium (Be). Among these water is the most widespread moderator. The properties of heavy-water are actually better but its disadvantage is the very high price.

If we start to put together more and more uranium (of course mixed with some moderator), the expected chain reaction will not start for a good while. The reason for this is that in the case of a small amount of fissile material large part of the neutrons that originate in fission will leak out of the pile and therefore they will not be able to cause fission. By increasing the amount of fissile material there will be a moment when we reach the so called *critical mass* and a self sustaining reaction starts. For pure metal U235 (without moderator) the critical mass is somewhat less than 50 kg. Because of the very high density (19.2 g/cm$^3$) this amount of uranium makes a sphere whose diameter is about 17 cm.

As we have already mentioned, at the fission of U-235 on the average 2.5 neutrons are released but not all of these cause fission. The neutronic state can be characterized by the so called *multiplication factor (k)*. This is the ratio of the number of neutrons in a neutron "generation" and the number of neutons in the preceding generation. (Neutron generation is the time that elapses between the birth of a neutron and the birth of neutrons from the subsequent fission which is caused by the given neutron. Its tipical value is about $10^{-9\ s}$ [i.e. 1 nanosecond] for pure U-235.)

If the value of the multiplication factor is 1, we talk about a *critical* reactor. In this case the number of neutrons in the system is constant, i.e. they cause the same number of fissions in every second. If k<1, the reactor is *subcritical* and the number of neutrons is going down. On the other hand, if k>1, the system is *supercritical*.

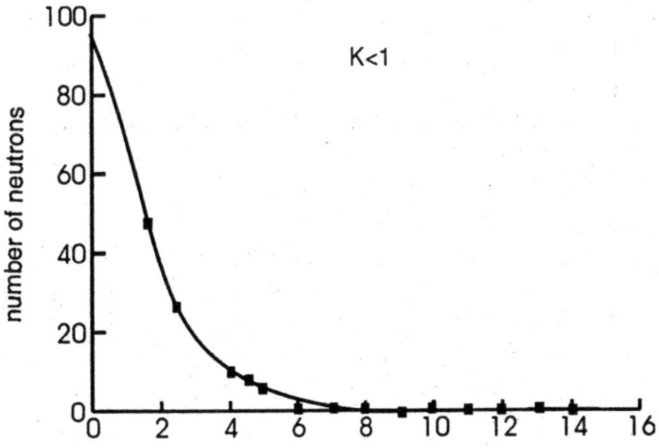

Very often the notion of *reactivity* is used instead of the multiplication factor. The definition of reactivity is ρ=(k-1)/k.

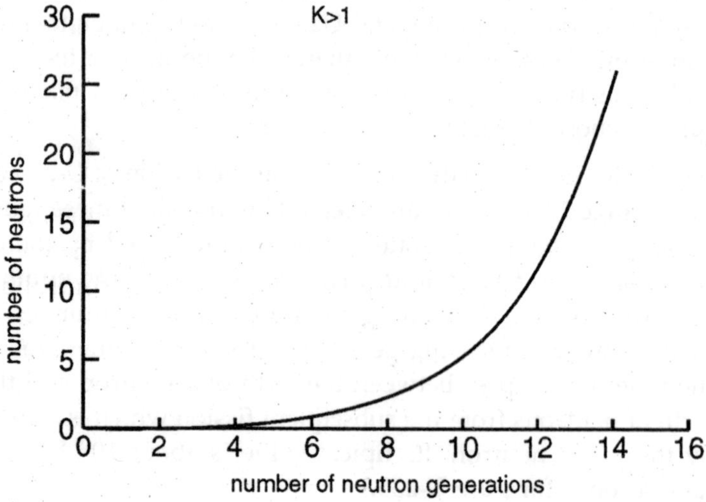


Correspondingly, for a subcritical reactor ρ<0, for a supercritical ρ>0, and for a critical ρ=0.

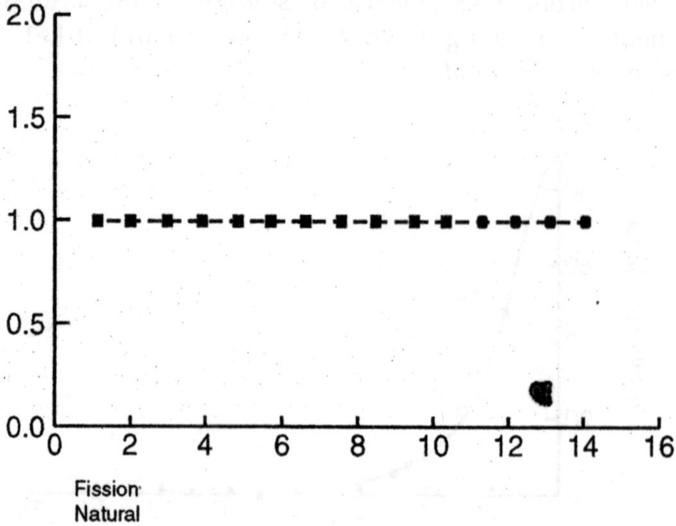

Obviously, the number of neutrons present in the reactor must be regulated or controlled since this determines the rate of fission and thus the energy released per second. In order to *control* the chain reaction one should use materials which tend to capture neutrons at a high probability. The most widespread neutron absorbers are cadmium (Cd) and boron (B).

The so called *control rods* are important tools of control and can be found in every reactor. These are made of neutron absorbing materials and can be moved between the fuel assemblies. If one wishes to decrease the power of the reactor it is sufficient to push a control rod a little inner. The control rods are particularly useful for the short term control and stopping the reactor. For long term regulation usually boric acid dissolved in the coolant is used.

The majority of the emitted neutrons leave the place of fission within an extremely short time. However, about 0.64percent of them (in the case of U-235) only leave the fission product significantly later. These are the so called delayed neutrons, which play a vital role in the controllability of reactors.

Change in the power results in changes in temperatures (fuel, coolant etc.). The temperature changes in turn modify the reactivity through different nuclear physics processes, which has a well defined influence on the number of neutrons ready to induce fission and thus on power. When the power rises, the temperature of fuel will rise too. However, the higher the temperature, the higher the ability of U-238 to absorb neutrons. Therefore, due to the presence of more neutron absorbing material reactivity is lower and the number of neutrons and accordingly power will decrease. Finally we may say that a chnge in power caused a change in reactivity and this in turn had an effect on power. If the change in reactivity is such that the processes caused are against the originating ones, we talk about negative feedback. The feedback was negative: the power increase induced a negative change in reactivity.

In terms of reactor safety it is an obvious intention that all feedbacks should be negative. In this case any power increase would cause a drop in reactivity and this will make the reactor sub-critical sooner or later, without external interaction. In such cases we say that the reactor has a so called inherent safety. Such inherent safety is characteristic for all PWRs and among them the reactors of the Paks NPP.

As a counter-example for inherent safety we may mention the RBMK type reactors, because in these positive feedbaks can occur in certain operational states. This contributed largely, among other factors, to the accident of the Chernobyl power plant.

**Advantages**

- Relatively little fuel is needed and the fuel is relatively inexpensive and available in trace amounts around the world.
- Fission is not believed to contribute to global warming or other pollution effects associated with fossil fuel combustion

**Disadvantages**

- Possibility of nuclear meltdown from uncontrolled reaction—leads to nuclear fallout with potentially harmful effects on civilians
- Waste products can be used to manufacture weapons
- High initial cost because plant requires containment safeguards

# Chapter 12

# Accelerator

A particle accelerator is a device that uses electric fields to propel electrically-charged particles to high speeds and to contain them. An ordinary CRT television set is a simple form of accelerator. There are two basic types: linear (i.e. straight-line) accelerators and circular (i.e. circles) accelerators.

## USES OF PARTICLE ACCELERATORS

Beams of high-energy particles are useful for both fundamental and applied research in the sciences. For the most basic inquiries into the dynamics and structure of matter, space, and time, physicists seek the simplest kinds of interactions at the highest possible energies. These typically entail particle energies of many GeV or more, and the interactions of the simplest kinds of particles: leptons (e.g. electrons and positrons) and quarks for the matter, or photons and gluons for the field quanta. Since isolated quarks are experimentally unavailable due to colour confinement, the simplest available experiments involve the interactions of, first, leptons with each other, and second, of leptons with nucleons, which are composed of quarks and gluons. To study the collisions of quarks with each other, we resort to collisions of nucleons, which at high energy may be usefully considered as essentially 2-body interactions of the quarks and gluons of which they are composed. Thus elementary particle physicists tend to use machines creating beams of electrons, positrons, protons, and anti-protons, interacting with each other or with the simplest nuclei (eg, hydrogen or deuterium) at the highest possible energies, generally hundreds of GeV or more.

At a higher level of complexity, nuclear physicists and cosmologists may use beams of bare atomic nuclei, stripped of electrons, to investigate the structure, interactions, and properties of the nuclei themselves, and of condensed matter at extremely high temperatures and densities, such as might have occurred in the first moments of the Big Bang. These investigations often involve collisions of heavy nuclei—of atoms like iron or gold—at energies of several GeV per nucleon. At lower energies, beams of accelerated nuclei are also used in medicine, as for the treatment of cancer.

Besides being of fundamental interest, high energy electrons may be coaxed into emitting extremely bright and coherent beams of high energy photons—ultraviolet and X ray—via synchrotron radiation, which photons have numerous uses in the study of atomic structure, chemistry, condensed matter physics, biology, and technology. Thus there is a great demand for electron accelerators of moderate (GeV) energy and high intensity.

**High-energy Machines**

Linear high-energy accelerators use a linear array of plates (or drift tubes) to which an alternating high-energy field is applied. As the particles approach a plate they are accelerated towards it by an opposite polarity charge applied to the plate. As they pass through a hole in the plate, the polarity is switched so that the plate now repels them and they are now accelerated by it towards the next plate. Normally a stream of "bunches" of particles are accelerated, so a carefully controlled AC voltage is applied to each plate to continuously repeat this for each bunch.

In early particle accelerators a Cockcroft-Walton voltage multiplier was responsible for voltage multiplying. This piece of the accelerator helped in the development of the atomic bomb. Built in 1937 by Philips of Eindhoven it currently resides in the National Science Museum in London, England.

As the particles approach the speed of light the switching rate of the electric fields becomes so high that they operate at microwave frequencies, and so RF cavity resonators are used in higher energy machines instead of simple plates.

DC accelerator types capable of accelerating particles to speeds sufficient to cause nuclear reactions are Cockcroft-Walton generators or voltage multipliers, which convert AC to high voltage DC, or Van de Graaff generators that use static electricity carried by belts.

The largest and most powerful particle accelerators, such as the RHIC, the LHC (scheduled to start operation in 2008) and the Tevatron, are used for experimental particle physics. Particle accelerators can also produce proton beams, which can produce "proton-heavy" medical or research isotopes as opposed to the "neutron-heavy" ones made in fission reactors. An example of this type of machine is LANSCE at Los Alamos.

### Low-energy Machines

Everyday examples of particle accelerators are those found in television sets and X-ray generators. Low-energy accelerators such as cathode ray tubes and X-ray generators use a single pair of electrodes with a DC voltage of a few thousand volts between them. In an X-ray generator, the target itself is one of the electrodes. A low-energy particle accelerator called an ion implanter is used in the manufacture of integrated circuits.

## LINEAR PARTICLE ACCELERATORS

In a linear accelerator (linac), particles are accelerated in a straight line with a target of interest at one end. Linacs are very widely used - every cathode ray tube contains one. They are also used to provide an initial low-energy kick to particles before they are injected into circular accelerators. The longest linac in the world is the Stanford Linear Accelerator, SLAC, which is 3 km (2 miles) long. SLAC is an electron-positron collider.

Linear accelerators are also widely used in medicine, for radiotherapy and radiosurgery. Medical grade LINACs accelerate electrons using a klystron and a complex bending magnet arrangement which produces a beam of 6-30 million electron-volt (MeV) energy. The electrons can be used directly or they can be collided with a target to produce a beam of X-rays. The reliability, flexibility and accuracy of the radiation beam produced has largely supplanted the older use of Cobalt-60 therapy as a treatment tool.

### Tandem Electrostatic Accelerators

In a tandem accelerator, the negatively charged ion gains energy by attraction to the very high positive voltage at the geometric centre of the pressure vessel. When it arrives at the centre region known as the high voltage terminal, some electrons are stripped from the ion. The ion then becomes positive and accelerated away by the high positive voltage. Thus, this type of accelerator is called a 'tandem' accelerator. The accelerator has two stages of acceleration, first pulling and then pushing the charged particles. An example of a tandem accelerator is ANTARES (Australian National Tandem Accelerator for Applied Research).

## CIRCULAR OR CYCLIC ACCELERATORS

In the circular accelerator, particles move in a circle until they reach sufficient energy. The particle track is typically bent into a circle using electromagnets. The advantage of circular accelerators over linear accelerators (*linacs*) is that the ring topology allows continuous acceleration, as the particle can transit indefinitely. Another advantage is that a circular accelerator is relatively smaller than a linear accelerator of comparable power (i.e. a linac would have to be extremely long to have the equivalent power of a circular accelerator).

Depending on the energy and the particle being accelerated, circular accelerators suffer a disadvantage in that the particles emit synchrotron radiation. When any charged particle is accelerated, it emits electromagnetic radiation and secondary emissions. As a particle traveling in a circle is always accelerating towards the center of the circle, it continuously radiates towards the tangent of the circle. This radiation is called synchrotron light and depends highly on the mass of the accelerating particle. For this reason, many high energy electron accelerators are linacs.

Since the special theory of relativity requires that matter always travels slower than the speed of light in a vacuum, in high-energy accelerators, as the energy increases the particle speed approaches the speed of light as a limit, never quite attained. Therefore particle physicists do not generally think in terms of speed, but rather in terms of a particle's energy or momentum, usually measured in electron volts (eV). An important principle for circular accelerators, and particle beams in general, is that the curvature of the particle trajectory is proportional to the particle charge and to the magnetic field, but inversely proportional to the (typically relativistic) momentum.

## Cyclotrons

The earliest circular accelerators were cyclotrons, invented in 1929 by Ernest O. Lawrence at the University of California, Berkeley. Cyclotrons have a single pair of hollow 'D'-shaped plates to accelerate the particles and a single large dipole magnet to bend their path into a circular orbit. It is a characteristic property of charged particles in a uniform and constant magnetic field B that they orbit with a constant period, at a frequency called the cyclotron frequency, so long as their speed is small compared to the speed of light c. This means that the accelerating D's of a cyclotron can be driven at a constant frequency by a radio frequency (RF) accelerating power source, as the beam spirals outwards continuously. The particles are injected in the centre of the magnet and are extracted at the outer edge at their maximum energy.

Cyclotrons reach an energy limit because of relativistic effects whereby the particles effectively become more massive, so that their cyclotron frequency drops out of synch with the accelerating RF. Therefore simple cyclotrons can accelerate protons only to an energy of around 15 million electron volts (15 MeV, corresponding to a speed of roughly 10% of c), because the protons get out of phase with the driving electric field. If accelerated further, the beam would continue to spiral outward to a larger radius but the particles would no longer gain enough speed to complete the larger circle in step with the accelerating RF. Cyclotrons are nevertheless still useful for lower energy applications.

## Synchrocyclotrons and Isochronous Cyclotrons

There are ways of modifying the classic cyclotron to increase the energy limit. This may be done in a continuous beam, constant frequency, machine by shaping the magnet poles so to increase magnetic field with radius. Then higher energy particles travel a shorter distance in each orbit than they would otherwise would, and can remain in phase with the accelerating field. Such machines are called isochronous cyclotrons. Their advantage is that they can deliver continuous beams of higher average intensity, which is useful for some applications. The main disadvantages are the size and cost of the large magnet needed, and the difficulty in achieving the higher field required at the outer edge.

Another possibility, the synchrocyclotron, accelerates the particles in bunches, in a constant B field, but reduces the RF accelerating field's frequency so as to keep the particles in step as they spiral outward. This approach suffers from low average beam intensity due to the bunching, and again from the need for a huge magnet of large radius and constant field over the larger orbit demanded by high energy.

## Betatrons

Another type of circular accelerator, invented in 1940 for accelerating electrons, is the Betatron. These machines, like sychrotrons, use a donut-shaped ring magnet with a cyclically increasing B field, but accelerate the particles by induction from the increasing magnetic field, as if they were the secondary winding in a transformer, due to the changing magnetic flux through the orbit. Achieving constant orbital radius while supplying the proper accelerating electric field requires that the magnetic flux linking the orbit be somewhat independent of the magnetic field on the orbit, bending the particles into a constant radius curve. These machines have in practice been limited by the large radiative losses suffered by the electrons moving at nearly the speed of light in a relatively small radius orbit.

## Synchrotrons

To reach still higher energies, with relativistic mass approaching or exceeding the rest mass of the particles (for protons, billions of electron volts GeV), it is necessary to use a synchrotron. This is an accelerator in which the particles are accelerated in a ring of constant radius. An immediate advantage over cyclotrons is that the magnetic field need only be present over the actual region of the particle orbits, which is very much narrower than the diameter of the ring. (The largest cyclotron built in the US had a 184 in dia magnet pole, whereas the diameter of the LEP and LHC is nearly 10 km. The aperture of the beam of the latter was of the order of centimeters.)

However, since the particle momentum increases during acceleration, it is necessary to turn up the magnetic field B in proportion to maintain constant curvature of the orbit. In consequence sychrotrons cannot accelerate particles continuously, as cyclotrons can, but must operate cyclically, supplying particles in bunches, which are delivered to a target or an external beam in beam "spills" typically every few seconds. Note that since high energy synchrotrons do most of their work on particles that are already traveling at nearly the speed of light c, the time to complete one orbit of the ring is nearly constant, as is the frequency of the RF power needed to drive the acceleration.

Note also a further point about modern synchrotrons: because the beam aperture is small and the magnetic field does not cover the entire area of the particle orbit as it does for a cyclotron, several necessary functions can be separated. Instead of one huge magnet, one has a line of hundreds of bending magnets, enclosing (or enclosed by) vacuum connecting pipes. The focusing of the beam is handled independently by specialized quadrupole magnets, while the acceleration itself is accomplished in separate RF sections, rather similar to short linear accelerators. Also, there is no necessity that cyclic machines be circular, but rather the beam pipe may have straight sections between magnets where beams may collide. be cooled, etc.

More complex modern synchrotrons such as the Tevatron, LEP, and LHC (still under construction) may deliver the particle bunches into storage rings of magnets with constant B, where they can continue to orbit for long periods for experimentation or further acceleration. The highest-energy machines such as the Tevatron and LHC are actually accelerator complexes, with a cascade of specialized elements in series, including linear accelerators for initial beam creation, one or more low energy synchrotrons to reach intermediate energy, storage rings where beams can be accumulated or "cooled" (reducing the magnet aperture required and permitting tighter focusing; see beam cooling), and a last large ring for final acceleration and experimentation.

## Electron Synchrotrons

Circular electron accelerators fell somewhat out of favour for particle physics around the time that SLAC was constructed, because their synchrotron losses were considered economically prohibitive and because their beam intensity was lower than for the unpulsed linear machines. The Cornell Electron Synchrotron, built at low cost in the late 1960s, was the first in a series of high-energy circular electron accelerators built for fundamental particle physics, culminating in the LEP at CERN.

A large number of electron synchrotrons have been built in the past two decades, specialized to be synchrotron light sources, of ultraviolet light and X rays.

## Storage Rings

For some applications, it is useful to store beams of high energy particles for some time (with modern high vacuum technology, up to many hours) without further acceleration. This is especially true for colliding beam accelerators, in which two beams moving in opposite directions are made to collide with each other, with a large gain in effective collision energy. Because relatively few collisions occur at each pass through the intersection point of the two beams, it is customary to first accelerate the beams to the desired energy, and then store them in storage rings, which are essentially synchrotron rings of magnets, with no significant RF power for acceleration.

## Synchrotron Radiation Sources

Some circular accelerators have been built to deliberately generate radiation (called synchrotron light) as X-rays also called synchrotron radiation, the Diamond Light Source being built at the Rutherford Appleton Laboratory in England or the Advanced Photon Source at Argonne National Laboratory in Illinois, USA. High-energy X-rays are useful for X-ray spectroscopy of proteins or X-ray absorption fine structure (XAFS).

Synchrotron radiation is more powerfully emitted by lighter particles, so these accelerators are invariably electron accelerators. Synchrotron radiation allows for better imaging as researched and developed at SLAC's SPEAR.

### *History*

Lawrence's first cyclotron was a mere 4 inches (100 mm) in diameter. Later he built a machine with a 60 in dia pole face, and planned one with a 184-inch dia, which was however taken over for WWII-related work connected with uranium isotope separation; after the war it continued in service for research and medicine over many years.

The first large proton synchrotron was the Cosmotron at Brookhaven National Laboratory, which accelerated protons to about 3 GeV. The Bevatron at Berkeley, completed in 1954, was specifically designed to accelerate protons to sufficient energy to create anti-protons, and verify the particle-antiparticle symmetry of nature, then only strongly suspected.

The AGS at Brookhaven was the first large synchrotron with alternating gradient, "strong focusing" magnets, which greatly reduced the required aperture of the beam, and correspondingly the size and cost of the bending magnets. The Proton Synchrotron, built at CERN, was the first major European particle accelerator and generally similar to the AGS.

The Fermilab Tevatron has a ring with a beam path of 4 miles (6 km). The largest circular accelerator ever built was the LEP synchrotron at CERN with a circumference 26.6 kilometers, which was an electron/positron collider. It has been dismantled and the underground tunnel is being reused for a proton/proton collider called the LHC, due to start operation in May 2008. The aborted Superconducting Supercollider (SSC) in Texas would have had a circumference of 87 km. Construction was started in 1991, but abandoned in 1993. Very large circular accelerators are invariably built in underground tunnels a few metres wide to minimize the disruption and cost of building such a structure on the surface, and to provide shielding against intense secondary radiations that may occur. These are extremely penetrating at high energies.

Current accelerators such as the Spallation Neutron Source, incorporate superconducting cryomodules. The Relativistic Heavy Ion Collider, and upcoming Large Hadron Collider also make use of superconducting magnets and RF cavity resonators to accelerate particles.

## TARGETS AND DETECTORS

The output of a particle accelerator can generally be directed towards multiple lines of experiments, one at a given time, by means of a deviating electromagnet. This makes it possible to operate multiple experiments without needing to move things around or shutting down the entire accelerator beam. Except for synchrotron radiation sources, the purpose of an accelerator is to generate high-energy particles for interaction with matter.

This is usually a fixed target, such as the phosphor coating on the back of the screen in the case of a television tube; a piece of uranium in an accelerator designed as a neutron source; or a tungsten target for an X-ray generator. In a linac, the target is simply fitted to the end of the accelerator. The particle track in a cyclotron is a spiral outwards from the centre of the circular machine, so the accelerated particles emerge from a fixed point as for a linear accelerator.

For synchrotrons, the situation is more complex. Particles are accelerated to the desired energy. Then, a fast acting dipole magnet is used to switch the particles out of the circular synchrotron tube and towards the target.

A variation commonly used for particle physics research is a collider, also called a *storage ring collider*. Two circular synchrotons are built in close proximity - usually on top of each other and using the same magnets (which are then of more complicated design to accommodate both beam tubes). Bunches of particles travel in opposite directions around the two accelerators and collide at intersections between them. This can increase the energy enormously; whereas in a fixed-target experiment the energy available to produce new particles is proportional to the square root of the beam energy, in a collider the available energy is linear.

## Higher Energies

At present the highest energy accelerators are all circular colliders, but it is likely that limits have been reached in respect of compensating for synchrotron radiation losses for electron accelerators, and the next generation will probably be linear accelerators 10 times the current length. An example of such a next generation electron accelerator is the 40 km long International Linear Collider, due to be constructed between 2015-2020.

As of 2005, it is believed that plasma wakefield acceleration in the form of electron-beam 'afterburners' and standalone laser pulsers will provide dramatic increases in efficiency within two to three decades. In plasma wakefield accelerators, the beam cavity is filled with a plasma (rather than vacuum). A short pulse of electrons or laser light either constitutes or immediately trails the particles that are being accelerated. The pulse disrupts the plasma, causing the charged particles in the plasma to integrate into and move toward the rear of the bunch of particles that are being accelerated. This process transfers energy to the particle bunch, accelerating it further, and continues as long as the pulse is coherent.

Energy gradients as steep as 200 GeV/m have been achieved over millimeter-scale distances using laser pulsers and gradients approaching 1 GeV/m are being produced on the multi-centimeter-scale with electron-beam systems, in contrast to a limit of about 0.1 GeV/m for radio-frequency acceleration alone. Existing electron accelerators such as SLAC could use electron-beam afterburners to greatly increase the energy of their particle beams, at the cost of beam intensity. Electron systems in general can provide tightly collimated, reliable beams; laser systems may offer more power and compactness. Thus, plasma wakefield accelerators could be used — if technical issues can be resolved — to both increase the maximum energy of the largest accelerators and to bring high energies into university laboratories and medical centres.

### Black Hole Production

In the next few decades, the possibility of black hole production at the highest energy accelerators may arise, if certain predictions of superstring theory are accurate. If they are produced, it is thought that black holes would evapourate extremely quickly via Hawking radiation. However, the existence of Hawking radiation is controversial. It is also thought that an analogy between colliders and cosmic rays demonstrates collider safety. If colliders can produce black holes, cosmic rays (and particularly ultra-high-energy cosmic rays) should have been producing them for eons, and they have yet to harm us.

## VAN DE GRAAFF GENERATOR

Van de Graaff generator. The large sphere acts as a capacitor to store the charge transferred up its supporting column. The small sphere (connected to the ground potential) will draw an arc from the larger when the air gap breakdown voltage is exceeded.

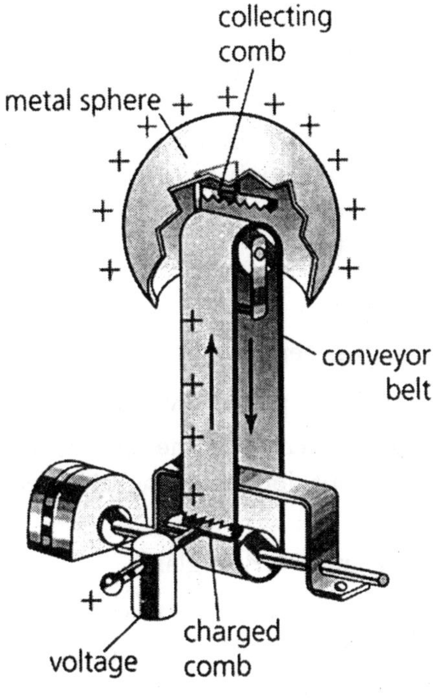


**Fig.** Van de Graaff generator

The Van de Graaff generator is an impressive electrostatic generator that is capable of producing enormously large static electric potentials. In fact, giant Van de Graaff generators can produce miliions of volts leading to awesome displays of corona and lightning. More modest "class room" sized Van de Graaff generators typically produce 100,000 V to 500,000 V. Originally the generator was built using a pure silk ribbon driven by a small motor. The charges on the ribbon were collected on a terminal that actually was a tin can. His early devices were capable of developing 80,000 volts. Of course, this device was limited by the edge effects of the tin can. However, his theory lead to Van de Graaff generators that have produced upwards to 10 million volts. Interestingly, Van de

Graaff may not have been the first to develop a belt driven electrostatic generator. In 1893, Von Busch presented a device having two pulleys and a horizontal belt with a charge collector comb and an insulated sphere. Even earlier than this, Rouland invented an electrostatic generator in 1785 using a continuous silk ribbon running between two horizontal pulleys with a collector tube at the center. Regardless of its "true" inventor, Van de Graaff generators have been instrumental in a number of fields, including particle acceleration, electrostatics, ESD, as well as providing an awesome display of the power of electricity.

## DESCRIPTION

Van de Graaff generator consist of:

1. Hollow metallic sphere (with positive charges)
2. Electrode connected to the sphere, a mesh in close proximity (but not contacting) the electrode and the belt
3. Upper roller (in acrylic glass)
4. Side of the belt with positive charges
5. Opposite side of the belt with negative charges
6. Lower roller (metal)
7. Lower electrode (ground)
8. Spherical device with negative charges, used to discharge the main sphere
9. Spark produced by the difference of potentials

A simple Van de Graaff generator consists of a belt of silk, or a similar flexible dielectric material, running over two pulleys, one of which is surrounded by a hollow metal sphere. Two electrodes, E1 and E2, in the form of comb-shaped rows of sharp metal points, are positioned respectively near to the bottom of the pulley and inside the sphere. E2 is connected to the sphere, and a high DC potential (with respect to earth) is applied to E1; a positive potential in this example.

The high voltage ionizes the air at the tip of E1, repelling (*spraying*) positive charges onto the belt, which then carries them up and inside the sphere. This positive charge induces a negative charge to the electrode E2 and a positive charge to the sphere (to which E2 is connected). The high potential difference ionizes the air inside the sphere, and negative charges are repelled from E2 and onto the belt, discharging it. As a result of the Faraday cage effect, positive charge on E2 migrates to the sphere regardless of the sphere's existing voltage. As the belt continues to move, a constant *charging current* travels via the belt, and the sphere continues to accumulate positive charge until the rate that charge is being lost (through leakage and corona discharges) equals the charging current. The larger the sphere and the farther it is from ground, the higher will be its final potential.

The other method for building Van de Graaff generators is to use the triboelectric effect. The two rollers for the belt are made of different materials, far from each other on the triboelectric series. When the belt comes into contact with one and is then separated, charge is transferred from the roller to the belt, and the roller becomes charged. When the belt comes into contact with the other roller and is then separated, charge is transferred from the belt to the roller, and that roller develops an opposite charge. The strong e-field from the rollers then induces a corona discharge at the tip of the pointed electrodes. The electrodes then "spray" a charge onto the belt which is opposite in polarity to the charge on the rollers. The remaining operation is otherwise the same as the voltage-injecting version above. This type of generator is easier to build for science fair or homemade projects, since it doesn't require a potentially dangerous high voltage source. The trade-off is that it cannot build up as high a voltage as the other type, and operation may become difficult under humid conditions (which can severely reduce triboelectric effects).

Since a Van de Graaff generator can supply the same small current at almost any level of electrical potential, it is an example of a nearly ideal current source. The maximum achievable potential is approximately equal to the sphere's radius multiplied by the e-field where corona discharges begin to form within the surrounding gas. A polished spherical electrode 30 cm in diameter immersed in air at STP (which has a breakdown voltage of about 30 kV/cm) could be expected to develop a maximum voltage of about 450 kV.

In Van de Graaff generators, electric charge is transported to the high-voltage terminal on a rapidly moving belt of insulating material driven by a pulley mounted on the grounded end of the structure; a second pulley is enclosed within the large, spherical high- voltage terminal, as shown at left. The belt is charged by a comb of sharp needles with the points close to the belt a short distance from the place at which it moves clear of the grounded pulley. The comb is connected to a power supply that raises its potential to a few tens of kilovolts. The gas near the needle points is ionized by the intense electric field, and in the resulting corona discharge the ions are driven to the surface of the belt. The motion of the belt carries the charge into the high-voltage terminal and transfers it to another comb of needles, from which it passes to the outer surface of the terminal. A carefully designed Van de Graaff generator insulated by pressurized gas can be charged to a potential of about 20 megavolts. An ion source within the terminal then produces positive particles that are accelerated as they move downward to ground potential through an evacuated tube.

In most constant-voltage accelerators, Van de Graaff generators are the source of high voltage, and most of the electrostatic proton accelerators still in use are two-stage tandem accelerators. These devices provide a beam with twice the energy that could be achieved by one application of the high voltage. An ion source yields a beam of protons, which are accelerated to a low energy by an auxiliary high-voltage supply. This beam passes through a region containing a gas at low pressure, where some of the protons are converted to negative hydrogen ions by the addition of two electrons. As the mixture of charged particles moves through a magnetic field, those with negative charge are deflected into the accelerator tube, and those with positive charge are deflected away. The beam of negative ions is then accelerated toward the positive high-voltage terminal. In this terminal, the particles pass through a thin carbon foil that strips off the two electrons, changing many of the negative ions back into positive ions (protons). These, now repelled by the positive terminal, are further accelerated through the second part of the tube. At the output end of the accelerator, the protons are magnetically separated, as before, from other particles in the beam and directed to the target.

## History

The fundamental idea for the friction machine as high-voltage supply, using electrostatic influence to charge rotating disk or belt can be traced back to the 17th century or even before.

The Van de Graaff generator was developed, starting in 1929, by physicist Robert J. Van de Graaff at Princeton University. The first model was demonstrated in October 1929. The first machine used a silk ribbon bought at a five and dime store as the charge transport belt. In 1931 a version able to produce 1,000,000 volts was described in a patent disclosure. This version had two 60 cm diameter charge accumulation spheres mounted on Pyrex glass columns 180 cm high; the apparatus cost only $90.

A more recent development is the tandem Van de Graaff accelerator, containing one or more Van de Graaff generators, in which negatively charged ions are accelerated through one potential difference before being stripped of two or more electrons, inside a high voltage terminal, and accelerated again.

One of Van de Graaff's accelerators used two charged domes of sufficient size that each of the domes had laboratories inside - one to provide the source of the accelerated beam, and the other to analyse the actual experiment. The power for the equipment inside the domes came from generators that ran off the belt, and several sessions came to a rather spectacular end when a pigeon would try to fly between the two domes - causing them to discharge (The accelerator was set up in an airplane hangar).

By the 1970s, up to 14 million volts could be achieved at the terminal of a tandem that used a tank of high pressure sulfur hexafluoride ($SF_6$) gas to prevent sparking by trapping electrons. This allowed the generation of heavy ion beams of several tens of megaelectronvolts, sufficient to study light ion direct nuclear reactions. The highest potential sustained by a Van de Graaff accelerator is 25.5 MV, achieved by the tandem at the Holifield Radioactive Ion Beam Facility at Oak Ridge National Laboratory.

A further development is the pelletron, where the rubber or fabric belt is replaced by a chain of short conductive rods connected by insulating links, and the air-ionizing electrodes are replaced by a grounded roller and inductive charging electrode. The chain can be operated at much higher velocity than a belt, and both the voltage and currents attainable are much higher than with a conventional Van de Graaff machine.

## Chapter 13

# Tools of Nuclear Science

Presently, the most commonly used tools of nuclear science are accelerators, reactors, detectors, and computers. The technological development of these devices has gone hand in hand with advances in nuclear science, sometimes leading and sometimes following closely behind.

## NUCLEAR REACTORS

Nuclear reactors created not only large amounts of plutonium needed for the weapons programmes, but a variety of other interesting and useful radioisotopes. They produced $^{60}Co$, in which the non-conservation of parity was first discovered, and a number of transuranic isotopes that are used to study the limits of the periodic table. Reactors also produce isotopes for commercial and medical purposes:

1. $^{241}Am$–used in smoke detectors,
2. $^{60}Co$–used in industry to inspect weld quality, also used in cancer therapy,
3. $^{99m}Tc$–used for medical diagnosis, and
4. $^{137}Cs$–also used for medical therapy.

Reactor neutrons have been used for material studies that involve their scattering from the crystal planes.

### Detectors

The interactions of alpha, beta, and gamma radiations with matter produce positively charged ions and electrons.

Radiation detectors are devices that measure this ionization and produce an observable output. Early detectors used photographic plates to detect "tracks" left by nuclear interactions. The cloud chambers, used to discover sub-nuclear particles, needed photographic recording and a tedious measurement of tracks from the photographs. Advances in electronics, particularly the invention of the transistor, allowed the development of electronic detectors. Scintillator-type detectors use vacuum tubes to perform the initial conversion of light to electrical pulses. The amplification and storing these data follow the advances in transistor electronics. Miniaturization in electronics has revitalized types of gas-filled detectors. These detectors were developed as "single element" detectors and now have been revived into "multiple element" detectors with more than one thousand elements. Advances in materials, particularly ultra-pure materials, and methods of fabrication have been critical to the creation of new and better detectors.

As the requirements for greater accuracy, efficiency, or sensitivity increases, so does the complexity of the detector and its operation. The following list presents some types of commonly used detectors and includes comments on each of them:

*Geiger Counter:* The detector most common to the public is the Geiger-Mueller counter, commonly called the Geiger counter. It uses a gas-filled tube with a central wire at high voltage to collect the ionization produced by incident radiation. It can detect alpha, beta, and gamma radiation although it cannot distinguish between them. Because of this and other limitations, it is best used for demonstrations or for radiation environments where only a rough estimate of the amount of radioactivity is needed.

*Scintillation detectors:* Scintillators are usually solids (although liquids or gases can be used) that give off light when radiation interacts with them. The light is converted to electrical pulses that are processed by electronics and computers. Examples are sodium iodide (NaI) and bismuth germanate (BGO). These materials are used for radiation monitoring, in research, and in medical imaging equipment.

*Solid state X-ray and gamma-ray detectors:* Silicon and germanium detectors, cooled to temperatures slightly above that of liquid nitrogen (77 K), are used for precise measurements of X-ray and gamma-ray energies and intensities. Silicon detectors are good for X-rays up to about 20 keV in energy. Germanium detectors can be used to measure energy over the range of >10 keV to a few MeV. Such detectors have applications in environmental radiation and trace element measurements. Germanium gamma ray detectors play the central role in nuclear high-spin physics, where gamma rays are used to measure the rotation of nuclei. Large gamma-ray detection systems, such as Gammasphere and Eurogam are made of these detectors.

*Low-energy charged particle detectors:* Silicon detectors, normally operated at room temperature, play a major role in the detection of low-energy charged particles. Singly, they can determine the energy of incident particles. Telescopes (combinations of two or more Si detectors) can be used to determine the charge (Z) and mass (A) of the particle. This type of detector is used in environmental applications to look for alpha-particle emitters (such as radium) in the environment.

*Neutron detectors:* Neutrons are much harder to detect because they are not charged. They are detected by nuclear interactions that produce secondary charged particles. Boron trifluoride ($BF_3$) counters make use of the $^{10}B(n,\alpha)^{7}Li$ reaction to detect neutrons. Often one uses a moderator, such as paraffin, to slow the neutrons and thus increase the detection efficiency. These detectors are used to monitor the neutron fluxes in the vicinity of a reactor or accelerator. Liquid scintillators can measure both neutrons and gamma rays. By carefully measuring the shape of the electronic signal, scientists can and distinguish between these two types of particles.

**Table. A partial List of Detectors used in Nuclear Science. Some Detectors can be Used only in a Limited Energy Range.**

| Particle Type | Detector Type | Features |
|---|---|---|
| *Charged* | | |
| protons, nuclei, electrons, or pions | Geiger-Müller counters | portable radioactivity detector |
| | gas ionization counters | gas-filled chamber in an electric field |
| | multiwire chambers | good position resolution |
| | semiconductor detectors | good energy resolution |
| | magnetic spectrometers | good momentum resolution |
| | scintillators and photomultipliers | good timing resolution |
| | Cerenkov detectors | good particle identification |
| *Neutral* | | |
| photons | scintillators and photomultipliers | good timing, moderate energy resolution |
| | germanium semiconductor crystals | good energy resolution |
| neutrons | liquid scintillator or $BF_3$ tubes | via fission, capture gamma rays, or proton collisions |
| neutrinos | Cerenkov detectors | via neutrino-electron interactions |
| | nuclear reactions | detect resultant radiation |

*Neutrino Detectors:* Neutrinos interact very weakly with matter and are therefore very hard to detect. Thus, neutrino detectors must be very large. The Sudbury Neutrino Observatory in Canada, was developed to understand the solar neutrino problem (too few neutrinos come out of the Sun than expected) and contains an active volume of 1000 tonnes (metric tons) of deuterium oxide (heavy water).

This is a Cerenkov counter in which the interaction of the neutrino with the heavy water produces an electron moving faster than the speed of light in the water. The moving electron generates a cone of light that can be observed with photomultiplier tubes. Information from these tubes will be used to determine the energy and direction of the incident neutrino.

*High-energy charged particle detectors:* As the energy increases, large and even more complex detection systems are needed, some involving thousands of individual detectors. These detectors typically involve the "tracking" of large numbers of particles as they pass through the detector. Large magnets are required to bend the paths of the charged particles. Multi-wire detection systems with nearly a quarter of a million channels of electronics provide information on these tracks. High speed computer systems process and store the data from these detectors. Similarly, powerful computer systems are needed to analyse these data so that a scientific discovery can be made.

The information that is presented in this section. It shows the different types of detectors that are suitable for measuring specific particles. When an experiment is designed, first a scientist chooses a particular detector based on the particles and their properties (such as energy, position, or time) that must be measured. Some detectors, such as scintillators, can make accurate time measurements but only a fair position determination. A scientist designs an experiment using an optimum choice of detector system. Cost is a major factor in modern detector design, especially for large systems consisting of a multitude of detectors and associated electronics.

### Computers

Beginning in the 1970s, computers played a role in nuclear science that developed from relatively minor to significant. Before this time, computers were used for calculations to develop and refine theories in nuclear science. As computers moved to being interfaced with detectors and accelerators, they became inseparable from the experiment. Indeed, the design of detectors for large experiments includes the integration of computer systems into each detector element. Computers are still used to calculate predictions of experiments based on various theories. Only the most powerful computer systems can generate simulations of the expected data from today's giant experiments. Similarly, only the most powerful computers can process the data that come from these experiments.

### Other Sciences and Technologies

While technology has been a driving force for nuclear science research, this field has similarly pushed the limits of technology. Likewise, advances in other scientific disciplines have been important to the progress in nuclear science. Development and advances in chemistry were essential to the discovery of most of the transuranic elements. This technology is still used to separate chemical species and allows studies of nuclei produced in accelerator or reactor experiments. Advances in solid state physics have produced larger and better silicon and germanium detectors for use in x-ray, gamma-ray, and particle spectroscopy. Advances in ultralow-temperature physics have produced superconducting magnets. They are used by the Michigan State University Cyclotron, by the superconducting radio frequency acceleration cavities at the Argonne National Laboratory's ATLAS accelerator, at the Jefferson Laboratory, and at the new RHIC collider.

## RADIOACTIVE WASTE

Radioactive wastes are waste types containing radioactive chemical elements that do not have a practical purpose. They are sometimes the products of nuclear processes, such as nuclear fission. However, other industries not directly connected to the nuclear industry can produce large quantities of radioactive waste. For instance, over the past 20 years it is estimated that just the oil-producing endeavors of the US have accumulated 8 million tons of radioactive wastes. The majority of radioactive waste is "low-level waste", meaning it has low levels of radioactivity per mass or volume. This type of waste often consists of used protective clothing, which is only slightly contaminated but still dangerous in case of radioactive contamination of a human body through ingestion, inhalation, absorption, or injection.

Many activities dealing with radioactive materials produce nuclear wastes, including civilian nuclear power programmes (nuclear powerplant operations and nuclear fuel-cycle activities), defence nuclear programmes, and industrial and institutional activities (scientific research, medical operations, and other industrial uses of radioisotopic sources or radiochemicals). To minimize the potential adverse health impacts to people during the entire lifetime of the radionuclides involved, nuclear waste must be carefully and properly managed. The scope of nuclear-waste management encompasses generation, processing (treatment and packaging), storage, transport, and disposal.

To avoid complexity in covering a wide variety of materials, nuclear waste is classified into several categories. According to the U.S. classification system, these are high-level waste (HLW), transuranic (TRU) waste, uranium mill tailings, and low-level waste (LLW). HLW includes the spent nuclear fuels withdrawn from nuclear power reactors following utilization for power generation and the highly radioactive liquid/solid waste resulting from the reprocessing of spent fuel. LLW includes all radioactive wastes other than spent fuel, HLW, TRU waste, and uranium mill tailings. HLW and LLW represent the two most important types of nuclear waste. More than 99% of the total radioactivity in nuclear waste is contained in HLW, while, in terms of volume, LLW takes up the biggest share (about 85% of the entire nuclear waste generated).

Materials issues are important during the entire process of nuclear-waste management— performance of the materials used in nuclear waste management determines its safety/ hazards. Since the safety of nuclear-waste management relies mainly on the immobilization of radioactive constituents and long-term isolation of these from the biosphere, materials issues with the immobilization and long-term isolation of nuclear waste are particularly important. As the requirements for materials performance are dependent upon the type of nuclear waste, this article concentrates on LLW and HLW.

## Radioactive Waste Management

Like all industries, the thermal generation of electricity produces wastes. Whatever fuel is used, these wastes must be managed in ways which safeguard human health and minimise their impact on the environment. Nuclear power is the only energy industry which takes full responsibility for all its wastes, and costs this into the product.

Nuclear power is characterised by the very large amount of energy available from a very small amount of fuel. The amount of waste is also relatively small. However, much of the waste is radioactive and therefore must be carefully managed as hazardous waste. Since the radioactive wastes are essentially created in a nuclear power reactor, it is accepted that they are the responsibility of the country which uses uranium to generate power. There is no moral or legal basis for the responsibility to be elsewhere.

Radioactivity arises naturally from the decay of particular forms of some elements, called isotopes. Some isotopes are radioactive, most are not, though in this publication we concentrate on the former.

There are three kinds of radiation to consider: alpha, beta and gamma. A fourth kind, neutron radiation, generally only occurs inside a nuclear reactor.

Different types of radiation require different forms of protection:

- Alpha radiation cannot penetrate the skin and can be blocked out by a sheet of paper, but is dangerous in the lung.
- Beta radiation can penetrate into the body but can be blocked out by a sheet of aluminium foil.
- Gamma radiation can go right through the body and requires several centimetres of lead or concrete, or a metre or so of water, to block it.

All of these kinds of radiation are, at low levels, naturally part of our environment. Any or all of them may be present in any classification of waste.

Radioactive wastes comprise a variety of materials requiring different types of management to protect people and the environment. They are normally classified as low-level, medium-level or high-level wastes, according to the amount and types of radioactivity in them. Another factor in managing wastes is the time that they are likely to remain hazardous. This depends on the kinds of radioactive isotopes in them, and particularly the half lives characteristic of each of those isotopes. The half life is the time it takes for a given radioactive isotope to lose half of its radioactivity. After four half lives the level of radioactivity is 1/16th of the original and after eight half lives 1/256th.

The various radioactive isotopes have half lives ranging from fractions of a second to minutes, hours or days, through to billions of years. Radioactivity decreases with time as these isotopes decay into stable, non-radioactive ones. The rate of decay of an isotope is inversely proportional to its half life; a short half life means that it decays rapidly. Hence, for each kind of radiation, the higher the intensity of radioactivity in a given amount of material, the shorter the half lives involved.

Three general principles are employed in the management of radioactive wastes:

- Concentrate-and-contain
- Dilute-and-disperse
- Delay-and-decay.

The first two are also used in the management of non-radioactive wastes. The waste is either concentrated and then isolated, or it is diluted to acceptable levels and then discharged to the environment. Delay-and-decay however is unique to radioactive waste management; it means that the waste is stored and its radioactivity is allowed to decrease naturally through decay of the radioisotopes in it.

### Types of Radioactive Waste (radwaste)

Low-level Waste is generated from hospitals, laboratories and industry, as well as the nuclear fuel cycle. It comprises paper, rags, tools, clothing, filters etc. which contain small amounts of mostly short-lived radioactivity. It is not dangerous to handle, but must be disposed of more carefully than normal garbage. Usually it is buried in shallow landfill sites. To reduce its volume, it is often compacted or incinerated (in a closed container) before disposal. Worldwide it comprises 90% of the volume but only 1% of the radioactivity of all radwaste.

Intermediate-level Wastecontains higher amounts of radioactivity and may require special shielding. It typically comprises resins, chemical sludges and reactor components, as well as contaminated materials from reactor decommissioning. Worldwide it makes up 7% of the volume and has 4% of the radioactivity of all radwaste. It may be solidified in concrete or bitumen for disposal. Generally short-lived waste (mainly from reactors) is buried, but long-lived waste (from reprocessing nuclear fuel) will be disposed of deep underground.

High-level Waste may be the used fuel itself, or the principal waste from reprocessing this. While only 3% of the volume of all radwaste, it holds 95% of the radioactivity. It contains the highly-radioactive fission products and some heavy elements with long-lived radioactivity. It generates a considerable amount of heat and requires cooling, as well as special shielding during handling and transport. If the used fuel is reprocessed, the separated waste is vitrified by incorporating it into borosilicate (Pyrex) glass which is sealed inside stainless steel canisters for eventual disposal deep underground.

On the other hand, if used reactor fuel is not reprocessed, all the highly-radioactive isotopes remain in it, and so the whole fuel assemblies are treated as high-level waste. This used fuel takes up about nine times the volume of equivalent vitrified high-level waste which results from reprocessing and which is encapsulated ready for disposal. Both high-level waste and used fuel are very radioactive and people handling them must be shielded from their radiation. Such materials are shipped in special containers which prevent the radiation leaking out and which will not rupture in an accident.

Whether used fuel is reprocessed or not, the volume of high-level waste is modest, - about 3 cubic metres per year of vitrified waste or 25-30 tonnes of used fuel for a typical large nuclear reactor. The relatively small amount involved allows it to be effectively and economically isolated.

## Radioactive Materials in the Natural Environment

Naturally-occurring radioactive materials are widespread throughout the environment, although concentrations are very low and they are not normally harmful. Soil naturally contains a variety of radioactive materials - uranium, thorium, radium and the radioactive gas radon which is continually escaping to the atmosphere. Many parts of the earth's crust are more radioactive than the low-level waste described above. Radiation is not something which arises just from using uranium to produce electricity, although the mining and milling of uranium and some other ores brings these radioactive materials into closer contact with people, and in the case of radon and its daughter products, speeds up their release to the atmosphere.

## Wastes from the Nuclear Fuel Cycle

Radioactive wastes occur at all stages of the nuclear fuel cycle - the process of producing electricity from nuclear materials. The fuel cycle comprises the mining and milling of the uranium ore, its processing and fabrication into nuclear fuel, its use in the reactor, the treatment of the used fuel taken from the reactor after use and finally, disposal of the wastes. The fuel cycle is often considered as two parts - the "front end" which stretches from mining through to the use of uranium in the reactor - and the "back end" which covers the removal of used fuel from the reactor and its subsequent treatment and disposal. This is where radioactive wastes are a major issue.

## Residual Materials from the "Front End" of the Fuel Cycle

The annual fuel requirement for a 1000 MWe light water reactor is about 25 tonnes of enriched uranium oxide. This requires the mining and milling of perhaps 50,000 tonnes of ore to provide about 200 tonnes of uranium oxide concentrate ($U_3O_8$) from the mine. At uranium mines, dust is controlled to minimise inhalation of radioactive minerals, while radon gas concentrations are kept to a minimum by good ventilation and dispersion in large volumes of air. At the mill, dust is collected and fed back into the process, while radon gas is diluted and dispersed to the atmosphere in large volumes of air.

At the mine, residual ground rock from the milling operation contain most of the radioactive materials from the ore, such as radium. This material is discharged into tailings dams which retain the remaining solids and prevent any seepage of the liquid. The tailings contain about 70% of the radioactivity in the original ore. Eventually these tailings may be put back into the mine or they may be covered with rock and clay, then revegetated. In this case considerable care is taken to ensure their long-term stability and to avoid any environmental impact (which would be more from acid leaching or dust than from radioactivity as such).

The tailings are usually around ten times more radioactive than typical granites, such as used on city buildings. If someone were to live continuously on top of the Ranger tailings they would receive about double their normal radiation dose from the actual tailings (ie they would triple their received dose). With in situ leach (ISL) mining, dissolved materials other than uranium are simply returned underground from where they came, as the water is recirculated.Uranium oxide ($U_3O_8$) produced from the mining and milling of uranium ore is only mildly radioactive - most of the radioactivity in the original ore remains at the mine site in the tailings.

Turning uranium oxide concentrate into a useable fuel has no effect on levels of radioactivity and does not produce significant waste. First, the uranium oxide is converted into a gas, uranium hexafluoride ($UF_6$), as feedstock for the enrichment process. Then, during enrichment, every tonne of uranium hexafluoride becomes separated into about 130 kg of enriched $UF_6$ (about 3.5% U-235) and 870 kg of 'depleted' $UF_6$ (mostly U-238). The enriched $UF_6$ is finally converted into uranium dioxide ($UO_2$) powder and pressed into fuel pellets which are encased in zirconium alloy tubes to form fuel rods.

Depleted uranium has few uses, though with a high density (specific gravity of 18.7) it has found uses in the keels of yachts, aircraft control surface counterweights, anti-tank ammunition and radiation shielding. It is also a potential energy source for particular (fast neutron) reactors.

## Wastes from the "Back End" of the Fuel Cycle

It is when uranium is used in the reactor that significant quantities of highly radioactive wastes are created. When the uranium-235 atom is split it forms fission products, which are very radioactive and make up the main portion of nuclear wastes retained in the fuel rods. There is also a relatively small amount of radioactivity induced in the reactor components by neutron irradiation.

About 25 tonnes of used fuel is taken each year from the core of a 1000 MWe nuclear reactor. This fuel can be regarded entirely as waste (as, for 40% of the world's output, in USA and Canada), or it can be reprocessed (as in Europe and Japan). Whichever option is chosen, the used fuel is first stored for several years under water in cooling ponds at the reactor site.

The concrete ponds and the water covering the fuel assemblies provide radiation protection, while removing the heat generated during radioactive decay.

**Fig.** Storage pond for used fuel at UK reprocessing plant

The costs of dealing with this high-level waste are built into electricity tariffs. For instance, in the USA, consumers pay 0.1 cents per kilowatt-hour, which utilities pay into a special fund. So far more than US$ 26 billion has been collected thus.

## REPROCESSING

If the used fuel is later reprocessed, it is dissolved and separated chemically into uranium, plutonium and high-level waste solutions. About 97% of the used fuel can be recycled leaving only 3% as high-level waste. The recyclable portion is mostly uranium depleted to less than 1% U-235, with some plutonium, which is most valuable. Arising from a year's operation of a typical 1000 MWe nuclear reactor, about 230 kilograms of plutonium (1% of the spent fuel) is separated in reprocessing. This can be used in fresh mixed oxide (MOX) fuel (but not weapons, due its composition). MOX fuel fabrication occurs in Europe, with some 25 years of operating experience. The main plant is in France, and started up in 1995.

Japan's slightly smaller plant is due to start up in 2012. Across Europe, over 35 reactors are licensed to load 20-50% of their cores with MOX fuel. The separated high-level wastes - about 3% of the typical reactor's used fuel - amounts to 700 kg per year and it needs to be isolated from the environment for a very long time. Major commercial reprocessing plants are operating in France and UK, with capacity of almost 5000 tonnes of spent fuel per year, - equivalent to at least one third of the world's annual output. A total of over 90,000 tonnes of spent fuel has been reprocessed at these over 40 years.

## Immobilising High-level Waste

Solidification processes have been developed in several countries over the past fifty years. Liquid high-level wastes are evapourated to solids, mixed with glass-forming materials, melted and poured into robust stainless steel canisters which are then sealed by welding. Borosilicate glass from the first waste vitrification plant in UK in the 1960s. This block contains material chemically identical to high-level waste from reprocessing. A piece this size would contain the total high-level waste arising from nuclear electricity generation for one person throughout a normal lifetime.The vitrified waste from the operation of a 1000 MWe reactor for one year would fill about twelve canisters, each 1.3m high and 0.4m diameter and holding 400 kg of glass. Commercial vitrification plants in Europe produce about 1000 tonnes per year of such vitrified waste (2500 canisters) and some have been operating for more than 20 years.

A more sophisticated method of immobilising high-level radioactive wastes has been developed in Australia. Called 'SYNROC' (synthetic rock), the radioactive wastes are incorporated in the crystal lattices of the naturally-stable minerals in a synthetic rock. In other words, copying what happens in nature. This process is now being tested in USA.

## WASTE DISPOSAL

Final disposal of high-level waste is delayed for 40-50 years to allow its radioactivity to decay, after which less than one thousandth of its initial radioactivity remains, and it is much easier to handle. Hence canisters of vitrified waste, or used fuel assemblies, are stored under water in special ponds, or in dry concrete structures or casks for at least this length of time.

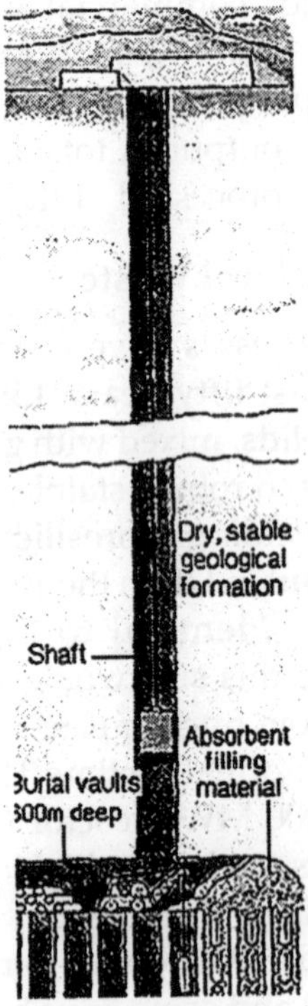


**Fig.** Nuclear Waste disposal

The ultimate disposal of vitrified wastes, or of used fuel assemblies without reprocessing, requires their isolation from the environment for long periods. The most favoured method is burial in dry, stable geological formations some 500 metres deep. Several countries are investigating sites that would be technically and publicly acceptable. The USA is pushing ahead with a repository site in Nevada for all the nation's used fuel.

One purpose-built deep geological repository for long-lived nuclear waste (though only from defence applications) is already operating in New Mexico.

After being buried for about 1000 years most of the radioactivity will have decayed. The amount of radioactivity then remaining would be similar to that of the naturally-occurring uranium ore from which it originated, though it would be more concentrated.

## Layers of Protection

To ensure that no significant environmental releases occur over a long perio after disposal, a 'multiple barrier' disposal concept is used to immobilise the radioactive elements in high-level (and some intermediate-level) wastes and to isolate them from the biosphere. The principal barriers are:

- Immobilise waste in an insoluble matrix, eg borosilicate glass, Synroc (or leave them as uranium oxide fuel pellets - a ceramic)
- Seal inside a corrosion-resistant container, eg stainless steel
- Surround containers with bentonite clay to inhibit any groundwater movement if the repository is likely to be wet,
- Locate deep underground in a stable rock structure

For any of the radioactivity to reach human populations or the environment, all of these barriers would need to be breached before the radioactivity decayed.

## A Natural Precedent

We have an example in nature to suggest that final disposal of high-level wastes underground is safe. Two billion years ago at Oklo in Gabon, West Africa, chain reactions started spontaneously in concentrated deposits of uranium ore. These natural nuclear reactors continued operating for hundreds of thousands of years forming plutonium and all the highly radioactive waste products created today in a nuclear power reactor. Despite the existence at the time of large quantities of water in the area, these materials stayed where they were formed and eventually decayed into non-radioactive elements. The evidence is there.

## Alternatives to Nuclear Electricity

No technology is absolutely safe or without environmental effects. We should therefore compare the production of electricity from nuclear energy with the other options available to us. Burning coal in power stations is still the major source of electricity worldwide, followed by hydro, uranium and gas.

A 1000 MWe light water reactor uses about 25 tonnes of enriched uranium a year, requiring the mining of some 50,000 tonnes of uranium ore. By comparison, a 1000 MWe coal-fired power station requires the mining, transportation, storage and burning of about 3.2 million tonnes of black coal per year. This creates around 7 million tonnes of carbon dioxide not to mention sulfur dioxide, depending on the particular coal. Solid wastes from a coal-fired power station can be substantial and cause environmental and health damage. Many people are concerned about possible global warming through enhancement of the greenhouse effect. Much of this is due to steadily increasing carbon dioxide in the atmosphere over the past 150 years. Burning fossil fuels, particularly coal, for electricity contributes almost 10 billion tonnes of carbon dioxide to the air each year.

## NUCLEAR-WASTE GENERATION

The proper selection of materials or quality control of materials manufacturing can lead to a reduction in radioactivity in nuclear waste. Reducing the amount of nitrogen impurities in nuclear fuels, both in the sintered $UO_2$ pellets and Zircaloy (zirconium alloyed mainly with tin and iron) cladding tube material, will lead to the reduction of C-14 activity in the spent fuel HLW. Use of Inconel-690 over Inconel-600 as a steam-generator tube material in pressurized water reactors can reduce the radioactivity in LLW from nuclear power plants due to the lower content of cobalt in Inconel-690. The use of low-cobalt-impurity, iron-based, hard-facing alloys instead of the Stellite alloy can have a similar effect.

Spent nuclear fuels can be reprocessed for recycling uranium or plutonium. This is achieved through the dissolution of fuels in nitric acids and selective solvent extraction using special organic solvent. HLW from the spent-fuel reprocessing can be treated with calcination or vitrification. In nuclear power plants, LLWs are produced through various treatment processes for radioactive liquids and gases prior to discharge to the environment; these processes include demineralization, filtration, or evapouration. Thus produced waste can be dewatered; compacted; supercompacted; incinerated; solidified; or manipulated through other special treatment processes, such as vitrification, thermal decomposition (molten-metal treatment or steam reforming), or chemical decomposition (supercritical water oxidation). The systems used for the processing and/or treatment of nuclear waste should be designed with careful consideration of materials in order to minimize the hazards during processing and to ensure the compatibility and the durability of the systems.

In transporting HLW, special transportation casks are designed to provide physical containment, radiation shielding, heat removal, criticality protection, and theft protection. Materials must be carefully selected to provide the required performance.

## Nuclear-waste Forms

A key consideration in nuclear-waste management is the development of a highly durable waste package (including the waste form and the surrounding container barriers) that ensures the long-term stability of materials and the isolation of radioactivity. Use of durable waste packages is also important in the interim storage of nuclear waste. In the waste package, the waste form represents the first and foremost barrier to the release of radionuclides from nuclear waste.

The functions of a waste form are to provide physically, chemically, and thermally stable form; immobilize the radio-active materials (slow release when contacted with water); and resist leaching, powdering, cracking, and other modes of degradation. The waste form into which the radionuclides are incorporated has a considerable impact on the manner and degree to which they are retained in the waste form.

In the case of LLW, typical waste forms include cement, polymers, glass, metals, sorbent materials, and various unsolidified waste. Cement is inexpensive, simple to fabricate, and stable in most environments. Various synthetic organic polymers (e.g., vinyl ester styrene) can be applicable at slightly higher fabrication cost when cement solidification is difficult (e.g., with ion-exchange resins). Glass, as the end-product of waste vitrification, is a much more stable waste form compared to cement or polymers. Vitrification can handle a wide variety of waste feeds with large volume reductions. However, some key radionuclides can be volatilized and released instead of being immobilized in the process.

In the United States, a large fraction of commercial LLW is disposed of without solidification. This is partly due to economic penalties in the disposalrate structure for increased waste volume caused by solidification. Vitrification is not widely practiced with its potentially higher cost involved. The U.S. Nuclear Regulatory Commission requirements for LLW form stress the structural stability of waste form, but do not necessarily promote the use of technology to enhance the long-term durability of waste forms. The majority of radionuclide inventory in disposed LLW currently exists in activated metals, dewatered ion-exchange resins, dewatered filters, and mixed trash (plastics and papers). In this case, the necessary stabilization or isolation of waste is mainly achieved by the use of waste containers.

In the case of HLW, activity contents (and their average half-life) are much greater than the LLW, thus requiring much more stringent control on waste-form performance. Selection of a waste form for HLW has been the subject of research for many years. The two primary HLW forms in the United States are spent fuel and borosilicate glass. Other major waste forms that have been considered include Synroc and tailored ceramics. Unreprocessed spent fuel can serve as a final waste form since the ceramic $UO_2$ matrix retains the nonvolatile fission products. The Zircaloy cladding, if intact, provides an additional barrier. Spent fuels are not vitrified in the United States.

Glass currently represents a smaller volume of HLW as compared to spent fuel in the United States.; however, it is the majority of waste in United Kingdom, France, Germany, Japan, and presumably Russia. The emphasis on the clean-up of the defence sites has also placed a premium on glass technology that is being used to immobilize contaminated materials and sites. The waste loading determines the volume, radiation dose, and thermal history of the waste form. The borosilicate glass is based on the $Na_2O$-$B_2O_3$-$SiO_2$ ternary system, with the waste loading ranging up to 30%. The oxide of boron is used to reduce the melting temperature without a large sacrifice in leachability. A wide range of fission products can be accommodated due to the geometrical flexibility afforded by disordered amorphous structure. Worldwide efforts have been made since the 1950s for the preparation of a practical, waste-immobilization scheme in this glass, including the building and operation of fully engineered smelters for the remote production of nuclear-waste glass. Thus, there is tremendous momentum behind borosilicate-glass technology in many countries.

Synroc is a titanium-based, polyphase, ceramic material made of specific natural minerals, including $TiO_2$, $ZrO_2$, $Al_2O_3$, BaO, and CaO. These minerals have the capacity to incorporate a wide range of radionuclides present in HLW into their crystal structures as solid solutions with approximately 20% waste loading. Tailored ceramics refer to multiphase, crystalline, ceramic nuclear-waste forms, which are similar to Synroc but vary depending on the compositions of the given waste (hence, tailored). Tailored ceramics are known to have slightly lower waste loading than Synroc (5– 15% waste).

The waste form is subject to various corrosive and degradation mechanisms that can decrease the effectiveness of the material to act as a barrier to nuclide migration.

Corrosion and dissolution, as well as radiation damage, can greatly reduce the desired material properties of the waste form. For the spent-fuel waste form, corrosion is a result of oxidation and hydriding of Zircaloy clad surrounding the spent-fuel pellets. As a result, the thermal conductivity of the clad decreases, increasing the thermal load on the waste form. Also, a decrease in the strength of the clad wall occurs due to the thinning of the clad.

Radiation hardening is also a key player in the degradation of spent fuel through creation of point, line, surface, and volumetric defects that inhibit dislocation movement. This can be very detrimental to spent-fuel integrity when the resulting loss of clad ductility is combined with effects such as fuel-pellet swelling or pellet-cladding mechanical interaction (PCI) of the clad with highly reactive fission products, such as iodine, cadmium, cesium, and molybdenum.

Other issues of importance in long-term repository performance assessment are the inhomogeneous distribution of fission products in the spent fuel and long-term dissolution of the $UO_2$ fuel matrix. Most of the release of soluble radionuclides occurs very early in the exposure to water from gap and grains, and their inventory is a function of fuel burn-up. The long-term release of most actinides is controlled by dissolution of the $UO_2$ matrix. Under reducing conditions, solubility controls the dissolution rate. Under oxidizing conditions (i.e., with water radiolysis), solubility becomes high and the presence of oxidizing species and complexing agents mainly control the kinetics of oxidative dissolution. While the integrity of the spent fuel cladding is of concern during dry storage, the cladding barrier is not counted in the repository performance assessment; this provides further conservatism in the repository-performance assessment. In addition, with the current trend to attain higher burn-ups, the performance of spent fuel as a barrier will be worse.

Glasses suffer from high-temperature thermal environment in a waste package. The extremely high temperatures in waste forms can cause devitrification and/or dissolution of glasses, which in turn result in the release of nuclides from the glass or crystal matrix. However, most (if not all) national waste-management programmes specify extended cooling periods for the spent fuel in order to reduce the temperature excursion within a repository during post-emplacement periods. Other concerns with the glass waste form are the radiation stability and long-term leaching through corrosion. Radiation interactions in glass lead to atomic displacements (by heavy particle radiations) or chemical effects (from beta or gamma rays) enhancing the corrosion rate through disordering or radiolytic processes. Atomic displacements lead to volume/density changes; stored energy change and/or its release; and associated crystallization, hardening, and fracture. The long-term release of glass constituents is controlled by the combined effects of the diffusive limited transport of dissolved species through the surface alteration layer and the crystallization of secondary phases formed on the surface of the glass.

## NUCLEAR-WASTE CONTAINTERS

Waste containers provide protective barriers against physical and chemical stresses during transportation, interim storage, and disposal. Since some of the radionuclides in LLWs are short-lived, LLWs are classified into different subcategories (depending upon the activity contents and half-life of the nuclides in the waste), and different stabilization requirements are used for different subcategories accordingly. In the case of HLW, separate casks are used for transportation, and special emphasis is given to the disposal containers to ensure long-term isolation of the waste. The key performance parameter is the resistance to environmental attack (chemical performance). Mechanical performance, thermal/ neutronic performance, compatibility with other materials, fabricability, and previous experience, as well as cost, are also taken into account.

Waste containers used in the burial of LLW include carbon-steel drums, liners, and boxes and high-integrity containers (HICs). They are placed in a disposal facility with either soil or cement backfills. Carbon-steel containers are inexpensive, but can undergo both uniform corrosion and pitting corrosion within the soil and cemented systems. The life-time of carbon-steel containers in a disposal system is expected to be short (few years or longer); thus, steel is used primarily for the disposal of short-lived nuclides. HICs represent a more durable LLW container and are used for the disposal of long-lived high-activity waste. HICs can be made from corrosion-resistant metal alloys, reinforced concrete, high-density polyethylene (HDPE), or polymer-coated metals. Carbon-steel drums, boxes, or HICs can also be used with concrete modules to improve the long-term integrity of the package.

The type of HIC expected to be widely used in the future is a combination of both an HDPE and a concrete overpack. This type of HIC is expected to fail eventually by degradation of the concrete casing and creep of the high-density polymers. HICs are required to have a minimum lifetime of 300 years by current Nuclear Regulatory Commission regulations.

Waste containers for HLW might include a metallic canister and an overpack.A canister is the immediate container surrounding the waste form. It serves as an aid to handling and transportation during processing and maintains geometric integrity for cooling; it also works as a short-term barrier against leaching and nuclide migration in disposal. Canisters cannot be considered a good barrier, especially if the hot glass material is put inside, since it is difficult to predict the thermal history of the metal. An overpack is a hermetically sealed, high-integrity protective barrier and provides the main protection for the desired long-term isolation during disposal. For the design of an HLW package, it is recommended to avoid dependence upon barriers in which geometric considerations may play a role in degradation modes (e.g., film growth, crevice corrosion, or flaw distribution) since it is difficult to predict the geometries of failures and nonuniform attacks.

HLW package materials will be subjected to harsh environments and various kinds of physical and chemical stresses. Long-term exposure of materials in a repository could result in significant alterations in materials during the service life. The presence of HLW inventory will lead to elevated temperatures and furnish high levels of radiation. The host media for the repository can be sources of oxygen, water, and other species that can be aggressive in altering the nature of the materials used for containment of the waste. Due to these stresses, various forms of degradation can be expected. One form of degradation is exposure to high-temperature gases that contain oxygen. These gases can cause oxidation of the materials, resulting in a loss of structural integrity, or could encourage future oxidation upon exposure of the container to groundwater. Another form of corrosion is due to the aqueous environment, including uniform corrosion, localized corrosion, galvanic corrosion, intergranular corrosion, and stress-corrosion cracking. Also, radiation effects, such as radiation hardening and embrittlement, enhanced diffusion, and enhanced creep rate, must be taken into account since all materials are susceptible to these phenomena.

Candidate materials for HLW canisters and overpacks are generally metals such as copper, iron, stainless steels, titanium alloys, and nickel-based alloys. Certain ceramics or graphitic materials may also be considered. Copper is one of the few metals found in its native state in the geological environment. Studies on native copper deposits and archaeological artifacts indicate very good environmental durability. However, it is known to be poor in brine as well as in radiation environment. Iron provides good predictability since much is known about the material. It is not very corrosion resistant, but is less prone to catastrophic failures. Both the natural occurrences and archaeological analog exhibit similar low iron corrosion rates. Carbon steel was considered as the structurally strong outer layer for corrosion allowance material in the current U.S. Department of Energy's (DOE's) HLW package design. Titanium alloys are mechanically strong and possess good corrosion resistance. However, they can experience brittle failure with the uptake of hydrogen. Nickel-based alloys, such as Incoloysand Hastelloys, are similar to titanium in that they are very corrosion resistant. They are easier to weld than titanium, but could be more expensive. Stainless steels have good mechanical properties and are very corrosion resistant, but catastrophic failures are possible through stress-corrosion cracking or intergranular corrosion. Ceramic materials, such as graphite and silicon carbide, have excellent corrosion resistance and are very much abundant, while mechanical strength is a problem with graphite.

The features of the current Yucca Mountain repository design are given by Bullen et al. For defence HLW, the vitrified waste would be contained inside stainless-steel canisters and put into the overpack. As the design process continues, DOE is evaluating various design options that might increase the ability of the engineered barrier system to contain waste and could reduce uncertainty.

**Nuclear-waste Package Materials**

The longevity of manufactured materials in the repository environment over such long periods of time is subject to significant uncertainty. At the same time, the prediction of material performance is essential in the development and use of waste packages (waste forms and waste containers). In the absence of a good mechanistic understanding of a material's performance and data that span a wide range of the expected performance and physicochemical conditions, extremely conservative assumptions need to be considered. Many of the performance predictions rely on data collected over a relatively limited range of test conditions; thus, extrapolation of these data requires good mechanistic understanding. Without proper data support, any benefits that the waste forms or container might provide could be ignored; hence, it is highly desirable to improve the predictability of the materials performance. This also requires demonstration of quality control of the product.

Various technical issues must be addressed in the assessment of the long-term performance of the waste package in a geologic repository. As all components of a waste package may be altered in time within the repository environment, the environment for a waste package (both internal and external) must be well characterized. A demonstrated understanding of factors that might affect long-term service behaviour is required for the characterization of materials for the waste-package components. These factors include variations in characteristics such as chemical composition, stress state, microstructure, fabrication or production history, and thermodynamic phase equilibria. Various interactions may be expected from gaseous or aqueous media that are in contact with the materials of the waste package. For metallic containers, various forms of corrosion that result from interactions with water and oxygen are important, as are the effects of hydrogen, which may result from radiolysis of water and vapour or galvanic coupling with borehole liner or container support structures. The environment may produce hydrostatic or lithostatic pressure, which may alter the stress state in waste-package components. Radiation will change the environment and create species with the potential for accelerated degradation of the waste-package components.

Microbial species, if they are present in significant quantities, have the potential for interactions with the waste-package materials. The service life of the waste package must be determined based on the consideration of these interactions between the environment and the waste-package components, including joints, seals, and welds. To ensure proper quality standards, materials and product specifications for each component of the waste package are necessary, including the construction and intermediate and final products. Quality control must be exercised for the fabrication of various waste-package components through the use of qualified personnel and equipment, procedures, and in-process product examinations. Inspection criteria are required to determine the acceptability of fabricated components for repository emplacement. Finally, in-situ monitoring needs to be implemented to evaluate the performance of the waste package with time up to its permanent closure. This requires development of advanced non-destructive techniques, such as magnetic resonance, automated ball indentation (ABI), and others, in addition to standard methods such as x-ray, ultrasound, and acoustic emission.

## GREENHOUSE EFFECT

A schematic representation of the exchanges of energy between outer space, the Earth's atmosphere, and the Earth surface. The ability of the atmosphere to capture and recycle energy emitted by the Earth surface is the defining characteristic of the greenhouse effect.

The greenhouse effect is the process in which the emission of infrared radiation by the atmosphere warms a planet's surface. The name comes from an incorrect analogy with the warming of air inside a greenhouse compared to the air outside the greenhouse. The greenhouse effect was discovered by Joseph Fourier in 1824 and first investigated quantitatively by Svante Arrhenius in 1896.

The Earth's average surface temperature of 15 °C (59 °F) is about 33 °C (59 °F) warmer than it would be without the greenhouse effect. Global warming, a recent warming of the Earth's lower atmosphere, is believed to be the result of an enhanced greenhouse effect due to increased concentrations of greenhouse gases in the atmosphere. In addition to the Earth, Mars and Venus have greenhouse effects.

### Basic Mechanism

Pattern of absorption bands generated by various greenhouse gases and their impact on both solar radiation and upgoing thermal radiation from the Earth's surface. Note that a greater quantity of upgoing radiation is absorbed, which contributes to the greenhouse effect.

The Earth receives energy from the Sun in the form of radiation. Most of the energy is in visible wavelengths and in infrared wavelengths that are near the visible range (often called "near infrared"). The Earth reflects about 30% of the incoming solar radiation. The remaining 70% is absorbed, warming the land, atmosphere and oceans.

For the Earth's temperature to be in steady state so that the Earth does not rapidly heat or cool, this absorbed solar radiation must be very closely balanced by energy radiated back to space in the infrared wavelengths. Since the intensity of infrared radiation increases with increasing temperature, one can think of the Earth's temperature as being determined by the infrared flux needed to balance the absorbed solar flux. The visible solar radiation mostly heats the surface, not the atmosphere, whereas most of the infrared radiation escaping to space is emitted from the upper atmosphere, not the surface. The infrared photons emitted by the surface are mostly absorbed in the atmosphere by greenhouse gases and clouds and do not escape directly to space.

The reason this warms the surface is most easily understood by starting with a simplified model of a purely radiative greenhouse effect that ignores energy transfer in the atmosphere by convection (sensible heat transport) and by the evapouration and condensation of water vapour (latent heat transport). In this purely radiative case, one can think of the atmosphere as emitting infrared radiation both upwards and downwards. The upward infrared flux emitted by the surface must balance not only the absorbed solar flux but also this downward infrared flux emitted by the atmosphere. The surface temperature will rise until it generates thermal radiation equivalent to the sum of the incoming solar and infrared radiation.

A more realistic picture taking into account the convective and latent heat fluxes is somewhat more complex. But the following simple model captures the essence. The starting point is to note that the opacity of the atmosphere to infrared radiation determines the height in the atmosphere from which most of the photons are emitted into space. If the atmosphere is more opaque, the typical photon escaping to space will be emitted from higher in the atmosphere, because one then has to go to higher altitudes to *see* out to space in the infrared. Since the emission of infrared radiation is a function of temperature, it is the temperature of the atmosphere at this emission level that is effectively determined by the requirement that the emitted flux balance the absorbed solar flux.

But the temperature of the atmosphere generally decreases with height above the surface, at a rate of roughly 6.5 °C per kilometer on average, until one reaches the stratosphere 10-15 km above the surface. (Most infrared photons escaping to space are emitted by the troposphere, the region bounded by the surface and the stratosphere, so we can ignore the stratosphere in this simple picture.) A very simple model, but one that proves to be remarkably useful, involves the assumption that this temperature profile is simply fixed, by the non-radiative energy fluxes. Given the temperature at the emission level of the infrared flux escaping to space, one then computes the surface temperature by increasing temperature at the rate of 6.5 °C per kilometer, the environmental lapse rate, until one reaches the surface. The more opaque the atmosphere, and the higher the emission level of the escaping infrared radiation, the warmer the surface, since one then needs to follow this lapse rate over a larger distance in the vertical. While less intuitive than the purely radiative greenhouse effect, this less familiar *radiative-convective* picture is the starting point for most discussions of the greenhouse effect in the climate modeling literature.

The term "greenhouse effect" is a source of confusion in that actual greenhouses do not warm by this mechanism. Popular discussions often imply incorrectly that they do; this error is sometimes made even in materials from scientific or governmental agencies.

## Greenhouse Gases

Quantum mechanics provides the basis for computing the interactions between molecules and radiation. Most of this interaction occurs when the frequency of the radiation closely matches that of the spectral lines of the molecule, determined by the quantization of the modes of vibration and rotation of the molecule. (The electronic excitations are generally not relevant for infrared radiation, as they require energy larger than that in an infrared photon.)

The width of a spectral line is an important element in understanding its importance for the absorption of radiation. In the Earth's atmosphere these spectral widths are primarily determined by "pressure broadening", which is the distortion of the spectrum due to the collision with another molecule. Most of the infrared absorption in the atmosphere can be thought of as occurring while two molecules are colliding. The absorption due to a photon interacting with a lone molecule is relatively small. This three-body aspect of the problem, one photon and two molecules, makes direct quantum mechanical computation for molecules of interest more challenging. Careful laboratory spectroscopic measurements, rather than *ab initio* quantum mechanical computations, provide the basis for most of the radiative transfer calculations used in studies of the atmosphere.

The molecules/atoms that constitute the bulk of the atmosphere: oxygen ($O_2$), nitrogen ($N_2$) and argon (Ar); do not interact with infrared radiation significantly. While the oxygen and nitrogen molecules can vibrate, because of their symmetry these vibrations do not create any transient charge separation. Without such a transient dipole moment, they can neither absorb nor emit infrared radiation. In the Earth's atmosphere, the dominant infrared absorbing gases are water vapour, carbon dioxide, and ozone ($O_3$). The same molecules are also the dominant infrared emitting molecules. $CO_2$ and $O_3$ have "floppy" vibration motions whose quantum states can be excited by collisions at energies encountered in the atmosphere. Carbon dioxide is a linear molecule, but it has an important vibrational mode in which the molecule bends with the carbon in the middle moving one way and the oxygens on the ends moving the other way, creating some charge separation, a dipole moment, thus carbon dioxide molecules can absorb IR radiation. Collisions will immediately transfer this energy to heating the surrounding gas. On the other hand, other $CO_2$ molecules will be vibrationally excited by collisions. Roughly 5% of $CO_2$ molecules are vibrationally excited at room temperature and it is this 5% that radiates. A substantial part of the greenhouse effect due to carbon dioxide exists because this vibration is easily excited by infrared radiation. $CO_2$ has two other vibrational modes. The symmetric stretch does not radiate, and the asymmetric stretch is at too high a frequency to be effectively excited by atmospheric temperature collisions, although it does contribute to absorption of IR radiation.

The vibrational modes of water are at too high energies to effectively radiate, but do absorb higher frequency IR radiation. Water vapour has a bent shape. It has a permanent dipole moment (the O atom end is electron rich, and the H atoms electron poor) which means that IR light can be emitted and absorbed during rotational transitions, and these transitions can also be produced by collisional energy transfer. Clouds are also very important infrared absorbers. Therefore, water has multiple effects on infrared radiation, through its vapour phase and through its condensed phases. Other absorbers of significance include methane, nitrous oxide and the chlorofluorocarbons.

Discussion of the relative importance of different infrared absorbers is confused by the overlap between the spectral lines due to different gases, widened by pressure broadening. As a result, the absorption due to one gas cannot be thought of as independent of the presence of other gases. One convenient approach is to remove the chosen constituent, leaving all other absorbers, and the temperatures, untouched, and monitoring the infrared radiation escaping to space. The reduction in infrared absorption is then a measure of the importance of that constituent. More precisely, define the greenhouse effect (GE) to be the difference between the infrared radiation that the surface would radiate to space if there were no atmosphere and the actual infrared radiation escaping to space. Then compute the percentage reduction in GE when a constituent is removed. The table below is computed by this method, using a particular 1-dimensional model of the atmosphere. More recent 3D computations lead to similar results.

| Gas removed | percent reduction in GE |
|---|---|
| $H_2O$ | 36% |
| $CO_2$ | 9% |
| $O_3$ | 3% |

By this particular measure, water vapour can be thought of as providing 36% of the greenhouse effect, and carbon dioxide 9%, but the effect of removal of both of these constituents will be greater than the total that each reduces the effect, in this case more than 45%. An additional proviso is that these numbers are computed holding the cloud distribution fixed. But removing water vapour from the atmosphere while holding clouds fixed is not likely to be physically relevant. In addition, the effects of a given gas are typically nonlinear in the amount of that gas, since the absorption by the gas at one level in the atmosphere can remove photons that would otherwise interact with the gas at another altitude. The kinds of estimates presented in the table, while often encountered in the controversies surrounding global warming, must be treated with caution. Different estimates found in different sources typically result from different definitions and do not reflect uncertainties in the underlying radiative transfer.

## Positive Feedback and Runaway Greenhouse Effect

When there is a loop of effects such as the concentration of a greenhouse gas itself being a function of temperature, there is a feedback. If the effect is to act in the same direction on temperature it is a positive feedback; and if in the opposite direction it is a negative feedback. Sometimes feedback effects can be on the same cause as the forcing but it can also be via another greenhouse gas or on other effects such as change in ice cover affecting the planet's albedo.

Positive feedbacks do not have to lead to a runaway effect. With radiation from the Earth increasing in proportion to the fourth power of temperature, the feedback effect has to be very strong to cause a runaway effect. An increase in temperature from greenhouse gases leading to increased water vapour which is a greenhouse gas causing further warming is a positive feedback. This cannot be a runaway effect or the runaway effect would have occurred long ago. Positive feedback effects are common and can always exist while runaway effects are much rarer and cannot be operating at all times.

If the effects from the second iteration of the loop of effects is larger than the effects of the first iteration of the loop this will lead to a self perpetuating effect. If this occurs and the feedback only ends after producing a major temperature increase, it is called a runaway greenhouse effect. A runaway feedback could also occur in the opposite direction leading to an ice age. Runaway feedbacks are bound to stop, since infinite temperatures are not observed. They are allowed to stop due to things like a reducing supply of a greenhouse gas or a phase change of the gas or ice cover reducing towards zero or increasing toward a large size that is difficult to increase.

According to the clathrate gun hypothesis a runaway greenhouse effect could be caused by liberation of methane gas from hydrates by global warming if there are sufficient hydrates close to unstable conditions. It has been speculated that the Permian-Triassic extinction event was caused by such a runaway effect. It is also thought that large quantities of methane could be released from the Siberian tundra as it begins to thaw, methane being 21-times more potent a greenhouse gas than carbon dioxide.

A runaway greenhouse effect involving $CO_2$ and water vapour may have occurred on Venus. On Venus today there is little water vapour in the atmosphere. If water vapour did contribute to the warmth of Venus at one time, this water is thought to have escaped to space. Venus is sufficiently strongly heated by the Sun that water vapour can rise much higher in the atmosphere and is split into hydrogen and oxygen by ultraviolet light. The hydrogen can then escape from the atmosphere and the oxygen recombines. Carbon dioxide, the dominant greenhouse gas in the current Venusian atmosphere, likely owes its larger concentration to the weakness of carbon recycling as compared to Earth, where the carbon dioxide emitted from volcanoes is efficiently subducted into the Earth by plate tectonics on geologic time scales.

### Anthropogenic Greenhouse Effect

$CO_2$ production from increased industrial activity (fossil fuel burning) and other human activities such as cement production and tropical deforestation has increased the $CO_2$ concentrations in the atmosphere. Measurements of carbon dioxide amounts from Mauna Loa observatory show that $CO_2$ has increased from about 313 ppm (parts per million) in 1960 to about 375 ppm in 2005. The current observed amount of $CO_2$ exceeds the geological record of $CO_2$ maxima (~300 ppm) from ice core data.

Because it is a greenhouse gas, elevated $CO_2$ levels will increase global mean temperature; based on an extensive review of the scientific literature, the Intergovernmental Panel on Climate Change concludes that "*most of the observed increase in globally averaged temperatures since the mid-20th century is very likely due to the observed increase in anthropogenic greenhouse gas concentrations*".

Over the past 800,000 years, ice core data shows unambiguously that carbon dioxide has varied from values as low as 180 parts per million (ppm) to the pre-industrial level of 270ppm. Certain paleoclimatologists consider variations in carbon dioxide to be a fundamental factor in controlling climate variations over this time scale.

## Real Greenhouses

The term 'greenhouse effect' originally came from the greenhouses used for gardening, but it is a misnomer since greenhouses operate differently. A greenhouse is built of glass. It heats up mainly because the Sun warms the ground inside it and this warms the air in the greenhouse. The air continues to heat because it is confined within the greenhouse, unlike the environment outside the greenhouse where warm air near the surface rises and mixes with cooler air aloft. This can be demonstrated by opening a small window near the roof of a greenhouse: the temperature will drop considerably. It has also been demonstrated experimentally a "greenhouse" built of rock salt (which is transparent to infrared radiation) heats up just as one built of glass does. Greenhouses thus work primarily by preventing *convection*; the atmospheric greenhouse effect however reduces *radiation loss*, not convection. It is quite common, however, to find sources that make the erroneous "greenhouse" analogy. Although the primary mechanism for warming greenhouses is the prevention of mixing with the free atmosphere, the radiative properties of the glazing can still be important to commercial growers. With the modern development of new plastic surfaces and glazings for greenhouses, this has permitted construction of greenhouses which selectively control radiation transmittance in order to better control the growing environment

## Chapter 14

# Black Body Radiation

### HEATED BODIES RADIATE

We shall now turn to another puzzle confronting physicists at the turn of the century — just how heated bodies radiate. There was a general understanding of the mechanism involved — heat was known to cause the molecules and atoms of a solid to vibrate, and it was known that the molecules and atoms were themselves complicated patterns of electrical charges. From the experiments of Hertz and others Maxwell's predictions that oscillating charges emitted electromagnetic radiation had been confirmed, at least for simple antennas. It was known from Maxwell's equations that this radiation traveled at the speed of light and from this it was realised that light itself, and the closely related infrared heat radiation, were actually electromagnetic waves. The picture, then, was that when a body was heated, the consequent vibrations on a molecular and atomic scale included some oscillating charges. If one assumed that Maxwell's theory of electromagnetic radiation, which worked well in the macroscopic world, was also valid at the microscopic scale (tenths of nanometers), then these oscillating charges would radiate, presumably giving off heat and light.

### RADIATION ABSORBED

What is meant by the phrase "black body" radiation? The point is that radiation from heated bodies depends to some extent on the body heated. Let us back up momentarily and consider how differently different materials *absorb* radiation. Some, like glass, seem to absorb light hardly at all-the light goes through. For a shiny metallic surface, the light isn't absorbed either, it gets reflected. For a black material like soot, light and heat are almost completely absorbed, and the material gets warm. How can we understand these different behaviours in terms of light as an electromagnetic wave interacting with charges in the material, causing them to oscillate and absorb energy from the radiation? In the case of glass, evidently this doesn't happen, at least not much. A full understanding of how this works needs quantum mechanics, but the general idea is as follows. There are charges-electrons-in glass that are able to oscillate in response to an applied external oscillating electric field, but these charges are tightly bound to atoms, and only oscillate at certain frequencies. It happens that for ordinary glass none of these frequencies correspond to those of visible light, so there is no resonance with a light wave, and hence little energy absorbed. Glass is opaque at some frequencies outside the visible range (in general, both in the infrared and the ultraviolet). These are the frequencies at which the electrical charge distribution in the atoms or bonds can naturally oscillate.

A piece of metal has electrons free to move through the entire solid. This is why metals can conduct electricity. It is also why they are shiny. These unattached electrons oscillate together with large amplitude in response to the electrical field of an incoming light wave. They themselves then radiate electromagnetically, just like a current in an antenna. This radiation from the oscillating electrons is the reflected light. In this situation, little of the incoming radiant energy is absorbed, it is just reradiated, that is, reflected.

Soot, like a metal, will conduct an electric current, although not nearly so well. There are unattached electrons, which can move through the whole solid, but they keep bumping into things-they have a short mean free path. When they bump, they cause vibration, like a pinball machine, so they give up energy into heat. Although the electrons in soot have a short mean free path compared to that in a good metal, they move very freely compared with electrons in atoms, so they can accelerate and pick up energy from the electric field in the light wave. They are therefore effective intermediaries in transferring energy from the light wave into heat.

## ABSORPTION AND EMISSION

Heated bodies radiate by processes just like the absorption described above operating in reverse. Thus, for soot heat causes the lattice to vibrate more vigorously, giving energy to the electrons (imagine them as balls in a pinball machine with strongly vibrating barriers, etc.) and since the electrons are charged they radiate away excess kinetic energy. On the other hand, the electrons in a metal have very long mean free paths, the lattice vibrations affect them much less, so they are less effective in radiating away heat. It is evident from considerations like this that good absorbers of radiation are also good emitters.

At sufficiently high temperatures, all bodies become good radiators. Items heated until they glow in a fire look much more similar than they do at room temperature. For a metal, this can be understood in terms of a shortening of the mean free path by the stronger vibrations of the lattice interfering with the electron's passage.

## OBSERVING THE BLACK BODY SPECTRUM

Any body at any temperature above absolute zero will radiate to some extent, but the intensity and frequency distribution of the radiation depends on the detailed structure of the body. To make any progress in understanding radiation, we must specify the details of the body radiating. The simplest possible model is to consider a body which is a perfect absorber, and therefore the best possible emitter. For obvious reasons, this is called a "black body".

A good model absorber is a small hole in the side of a box. Radiation impinging on the hole from outside enters the box and is absorbed as it is scattered around inside, so only a tiny fraction is re-emitted.

We shall, therefore, take a cubical box with a small hole in the side as our model to analyse emission of radiation. Fortunately, this is a system which can also be investigated experimentally fairly easily. A beam of radiation from the hole in the oven is passed through a diffraction grating and projected on to a screen, where it is separated out by wavelength (or equivalently frequency). A detector is moved up and down along the screen to find how much radiant energy is being emitted in each frequency range, say the energy between frequency $f$ and $f + df$ is $R_T(f)df$.

We can then plot $R_T(f)$ against $f$ for a given temperature $T$. It is found experimentally that for small $f$, $R_T(f)$ is proportional to $f^2$, a parabolic shape, but for increasing $f$ it falls below the parabola, peaks at $f_{max}$, then falls quite rapidly to zero as $f$ increases beyond $f_{max}$.

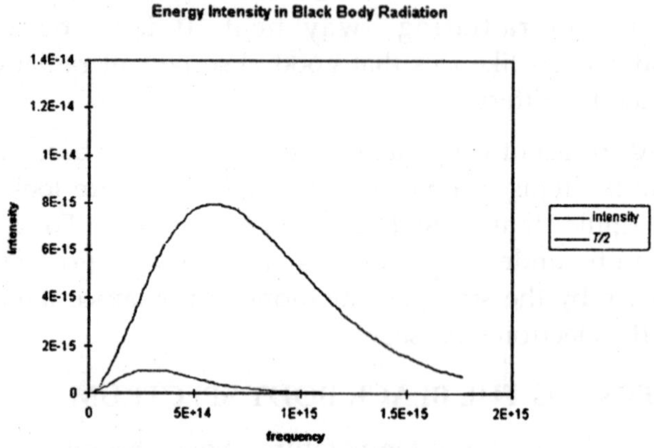


Suppose we now double the absolute temperature - how does that affect $R_T(f)$?

For those low frequencies where $R_T(f)$ was parabolic, doubling the temperature doubles the intensity of the radiation. However, it is found that at $2T$ the curve follows the (doubled) parabolic path much further - in fact, twice as far: $f_{max}(2T) = 2f_{max}(T)$.

The curve $R_{2T}(f)$, then, reaches *eight* times the height of $R_T(f)$. It also spreads over twice the lateral extent, so the area under the curve, corresponding to the total energy radiated, increases *sixteenfold* on doubling the temperature. This is Stefan's Law of Radiation: the power $P$ radiated from one square meter of black surface at temperature $T$:

$$P = \sigma T^4, \sigma = 5.67 \times 10^{-8} \text{ watts/sq.m/k}^4$$

Note that the graph above plots the energy intensity in the oven. This energy is being pumped out of the opening at speed $c$. When allowance is made for the fact that an opening of unit area is effectively less than unit area for radiation coming in at an angle, and half the energy is in waves moving in directions away from the hole, the actual rate of emission of energy through a hole of unit area is equal to $(1/4)c$ times the energy intensity inside. The blue curve in the graph above is for 10,000K. The energy intensity plotted on the $y$-axis is in joules/$m^3$ per unit frequency range, and it is evident on glancing at the graph that integrating to give the total intensity one will find something around 8 joules/$m^3$. To find the heat energy radiated per second per square meter of surface at 10,000K, we multiply this energy density by $c/4$, which gives around 600 Megawatts per square meter.

The upward shift in $f_{max}$ with $T$ is something everyone is familiar with - when an iron is heated in a fire, the first visible radiation (at around 900K) is deep red - the lowest frequency visible light. Further increase in $T$ causes the colour to change to orange then yellow, and white at a very high temperature, signifying that all the visible frequencies are being emitted roughly equally.

The change in $f_{max}$ is linear in the absolute temperature;

$$f_{max} \propto T$$

This is called *Wein's Displacement Law.*

For example, the sun's surface temperature is 5700K, and at that temperature much of the energy is radiated as visible light. This is no accident - evolution has adapted us to see most efficiently in the light most readily available.

This shift in the frequency at which radiant power is a maximum is very important for harnessing solar energy, such as in a greenhouse. We need glass which will allow the solar radiation in, but not let the heat radiation out. This is feasible because the two radiations are in very different frequency ranges - 5700K and, say, 300K - and there are materials transparent to light but opaque to infrared radiation. This is only possible because $f_{max}$ varies with temperature.

## UNDERSTANDING THE BLACK BODY SPECTRUM: THE ANALOGY WITH A GAS

Experimental investigations of the radiation intensity as a function of temperature and frequency, $R_T(f)$ above, give well-defined reproducible results, not very sensitive to the material of the oven, etc. For a small hole, the radiation emitted will be a representative sample of the radiation bouncing around inside the box.

In fact, the general shape of the function $R_T(f)$ was very reminiscent of the energy distribution in a gas in thermal equilibrium, as analysed by Maxwell and Boltzmann. In contrast to the molecules in a gas, the different wavelengths of radiation do not collide with each other in the middle of the oven. Nevertheless, energy can shift from one mode to the other through the intermediary of the oscillators in the wall - which is to say, just the atoms and molecules the wall is made of. An electromagnetic wave in the oven can set a charge oscillating in the wall, energy can be transferred to other charges in the wall, which will in general oscillate and radiate at different frequencies. Thus the different frequencies of radiation inside the oven will come to thermal equilibrium.

## COUNTING THE DEGREES OF FREEDOM

To make any quantitative progress in analyzing the radiation, we must have a clear picture of the degrees of freedom of this system - how many oscillators have frequencies in a given energy range? These oscillations are standing electromagnetic waves. The waves are contained in the oven, so the electric field intensity drops rapidly to zero on approaching and going into the walls, because inside the walls the electric energy will be rapidly dissipated by currents or polarization. In fact, then, the boundary condition at the walls is much like that for waves on a string fixed at both ends, where the wave amplitude goes to zero at the ends.

As a warm up exercise, consider the frequency distribution function of the possible different modes of vibration (that is, the different degrees of freedom) of a string stretched between two points a distance *a* apart. Possible values of the wavelength are:

$$\lambda = 2a, a, 2a/3...$$

so the frequency

$$f = c/\lambda = c/2a, 2(c/2a), 3(c/2a), ...$$

The allowed frequencies are thus equally spaced $c/2a$ apart. We define the spectral density by stating that the number of modes between $f$ and $f$ + delta $f$ is $N(f)$delta $f$ for delta $f$ small compared with the range of frequencies in the system, but large compared with the spacing between successive allowed frequencies. Evidently, in this simple one-dimensional example $N(f)$ is a constant equal to $2a/c$.

The amplitude of oscillation as a function of time has the form:

$$y = A \sin \frac{2\pi x}{\lambda} \sin 2\pi ft$$

This is more conveniently written:

$$y = A \sin kx \sin wt$$

where k = $2\pi/\lambda, w = 2\pi f$, so $w = ck$

Here is the *radial* frequency - how many *radians* per second, rather than cycles per second, the time dependent component of the wave moves through. Analogously, *k*, called the *wave number*, counts the radians of spatial change in unit length (such as one meter).

We are now ready to tackle the more complex problem of three-dimensional standing electromagnetic waves in a cubical oven. The first guess, just generalizing the expression above, would be:

$$\vec{E}(x,t) = \vec{E}_0 \sin k_x x \sin k_y y \sin k_z z \sin wt$$

This will satisfy the wave equation

$$\vec{\nabla}^2 \vec{E} - \frac{1}{c^2}\frac{\partial^2 \vec{E}}{\partial t^2} = 0$$

if $w^2 = c^2\left(k_x^2 + k_y^2 + k_z^2\right)$

Also the electric field will be zero at all the walls if we choose the $k$'s appropriately, that is, so that $\sin k_x a = 0$, etc.

However, this wave is not quite correct, because since there is no electric charge in the oven away from the walls, the divergence of the electric field must be identically zero away from the walls, a condition this wave does not satisfy. The divergence is:

$$\begin{aligned}\vec{E}(x,t) &= E_{0x}k_x \cos k_x x \sin k_y y \sin k_z z \sin wt + \\ &\quad E_{0y}k_y \cos k_x x \sin k_y y \sin k_z z \sin wt + \\ &\quad E_{0z}k_z \sin k_x x \sin k_y y \cos k_z z \sin wt\end{aligned}$$

and, for example, taking any $x$ for which $\sin k_x x = 0$, the second and third terms above are identically zero, but the first term certainly isn't.

This problem can be fixed by thinking more carefully about the boundary conditions at the walls. It is true that any component of the electric field *parallel* to the wall will be attenuated rapidly by currents or polarization in the wall. However, there could be an electric field *perpendicular* to the wall, because there could be surface charge on the wall. This means a possible solution is:

$$\vec{E}_x = E_{0x} \cos k_x x \sin k_y y \sin k_z z \sin wt$$

$$\vec{E}_y = E_{0y} \sin k_x x \cos k_y y \sin k_z z \sin wt$$

$$\vec{E}_z = E_{0z} \sin k_x x \sin k_y y \cos k_z z \sin wt$$

These components are deliberately chosen to give

$$div\vec{E} = \left(k_x E_{0x} + k_y E_{0y} + k_2 E_{02}\right)$$

$$\sin k_x x \sin k_y y \sin k_z z \sin wt$$

so choosing $\vec{k}$ so that $\vec{k} \cdot \vec{E}_0 = 0$, we find div $\vec{E} = 0$

This is telling us something we should have realised from the beginning - in the standing electromagnetic wave in the cavity, just as in a propagating wave, the electric field is perpendicular to the direction in which the wave is moving. In other words, it is a transverse wave, and in fact there are two independent polarizations, which we must remember to count when we find the total number of degrees of freedom.

It remains to find the allowed frequencies of vibration - the normal modes - of the electromagnetic radiation in the oven. The first step is to find the allowed values of the wave numbers $k_x$, $k_y$ and $k_z$. These are fixed by the boundary conditions $\sin k_x a = 0$, etc., so

$$k_x = \pi / a, 2\pi / a, 3\pi / a, \text{etc.}$$

Similarly, $k_y, k_z = \pi / a, 2\pi / a, \text{etc.},$

so the vector $\vec{k} = \left(k_x, k_y, k_z\right) = \dfrac{\pi}{a}(l, m, n)$

where($l$, $m$, $n$) are positive integers, like (1, 1, 1) or (5, 13, 4) - in other words, each possible standing electromagnetic wave in the oven corresponds to a point in the ($k_x$, $k_y$, $k_z$) space labeled by three positive integers. These are the *intersection points of a cubic lattice* in the space $k_x > 0$, $k_y > 0$, $k_z > 0$.

Each such point can be associated with the little lattice cube of volume for $\pi^3 / a^3$ which the point in question is the furthest corner of the little cube from the $k$-space origin. These little cubes stack together to fill all that part of $k$-space having $k_x > 0$, $k_y > 0$ and $k_z > 0$.

The frequency of vibration of the wave having wave numbers $(k_x, k_y, k_z)$ is

$$w = c\sqrt{k_x^2 + k_y^2 + k_z^2}$$

Recall that the experimental measurement of black body radiation from the oven detects the intensity of radiation in a given *frequency* range. It does *not* tell us the wave numbers of the radiating modes! Hence, to compare with these experimental results, we must choose a small frequency range and find how many possible sets of wave numbers correspond to modes of vibration having frequencies within that range.

Notice that a fixed value of the frequency corresponds to a spherical surface in $k$-space:

$$\frac{w^2}{c^2} = k_x^2 + k_y^2 + k_z^2, \text{ a sphere of radius w/c.}$$

Therefore, the number of possible modes of radiation in the oven having frequencies in the range $(w, w + \Delta w)$ is equal to the number of lattice-point wave number values $(k_x, k_y, k_z)$ between two spherical surfaces centred at the origin and having radii $(w/c, (w + \Delta w)/c)$, and only counting in the octant corresponding to $k_x > 0$, $k_y > 0$ and $k_z > 0$.

Now, as we argues above, each lattice point can be associated with a small cube of volume $\pi^3/a^3$. Assuming we choose $\Delta w >> a$, there will be many of these small cubes between the spherical surfaces, and the total number of lattice points between the spheres in the positive octant will be just the volume of the space between the spheres divided by the volume of one of these cubes.

So the number of possible modes of radiation in the oven having frequencies in the range $(w, w + \Delta w)$ is:

$$\frac{\frac{1}{8} \times 4\pi \frac{w^2}{c^2} \frac{\Delta w}{c}}{\frac{\pi^3}{a^3}} = \frac{1}{2} a^3 \frac{w^2 \Delta w}{c^3 \pi^2}$$

Putting this in terms of the frequency $f$ in cycles per second,, and inserting an extra factor of two for the two independent polarizations of the wave,we find the distribution function for modes in frequency is given by:

$$N(f)df = \frac{8V\pi f^2 df}{c^3}$$

where $V = a^3$ is the volume of the oven.

## EQUIPARTITION OF ENERGY

The point of the rather tedious calculation above, finding exactly how many different modes of vibration, or degrees of freedom, there are in the ovenfull of radiation, was to construct an argument parallel to that applied successfully to understand the properties of gases. The first conclusion would be that there is $kT$ of energy, on average in each wave mode not $\frac{1}{2}kT$, because the standing wave is like an oscillator with both kinetic and potential energy.

For low frequencies, this is exactly what is observed. The amount of radiation shining out of the oven at low frequencies is correctly predicted by counting modes, as above, and allocating $kT$ of energy to each mode.

## THE ULTRAVIOLET CATASTROPHE

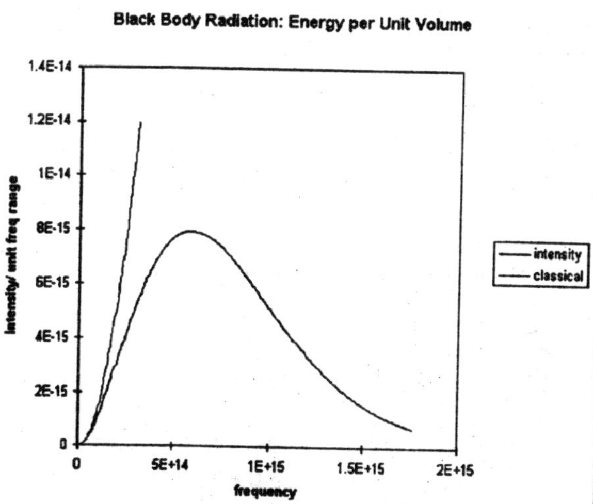

The problem is that as we go to higher frequencies, there are more and more possible degrees of freedom. The oven should be radiating huge amounts of energy in the blue and ultraviolet. But it isn't. For this system, *the equipartition of energy isn't working*!

The graph shows the prediction of radiation intensity from the classical equipartition compared with the experimental result. It is evident that the higher frequency modes are not getting their "fair share" $kT$ of energy, somehow they are getting frozen out, evidently those at really high frequencies are getting almost no share in the energy at all.

It is important at this point to go back to the curves earlier for two different temperatures, and ask the question: how is the sharing of energy among the modes affected by raising the temperature? We provide below a magnified version of the low frequency parts of the curves for T/2 and T, and show for each case the classical prediction based on the mode counting in the preceding section. Look at the curves and think about this before reading on!

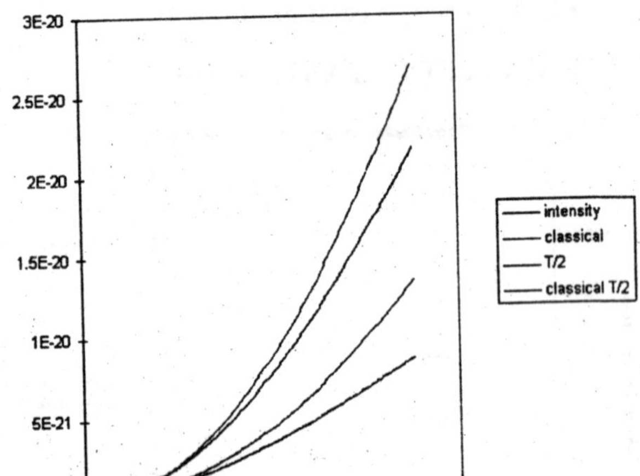


It is clear that for the lowest frequencies, doubling the temperature does indeed double the intensity of the radiation, exactly what we expect if the radiation is proportional to kT, the energy in each mode. (T = 300K here, by the way.) We can also see, though, that at the higher temperature, the observed radiation intensity stays closer to the classical prediction as we go to higher frequencies. This means the higher modes are less locked out of the energy sharing when we go to higher temperatures. For any given temperature, though, if we go to high enough frequencies, the intensity drops exponentially with further increase in frequency.

## PLANCK'S SUGGESTION

Planck suggested that the blackbody radiation curves could be understood if the oscillators in the walls of the oven were somehow constrained so that they could not continuously emit radiation, as Maxwell's equations would predict, but could only emit it in chunks - called *quanta* - of a definite magnitude. For higher frequencies, he suggested, the radiation had to be emitted in bigger chunks. This would explain the lack of radiation in the oven at these higher frequencies - assuming the oscillators in the walls had on average energy $kT$, it would be very unlikely that one oscillating charge would by random excitation have an energy of, say, $5kT$, so if its frequency of oscillation were high enough, it would almost never have enough energy to emit one quantum, so by Planck's theory, it wouldn't emit any radiation at all. On the other hand, if the temperature were increased fivefold, it would often have enough energy to emit a quantum.

Given, then, that bigger chunks of energy are emitted at the higher frequencies, just how does the size of a chunk depend on the frequency of the radiation? The clue is given by the Wein displacement formula, $f_{max} \alpha T$. Now, $f_{max}$ is the frequency at which the curve turns over and begins to descend, moving far from the classically predicted parabola, which of course continues upwards. Thus, doubling the temperature doubles the frequency range over which the curve approximates the classically predicted parabola. (If we stiffen the requirements for approximating the classical curve, for example by requiring it to be within 5per cent, say, then we can see from the graph of low energy modes above that doubling the temperature still doubles the range over which the curves remain close.)

The bottom line is that doubling the temperature apparently doubles the frequency range in which oscillators get approximately their full share $kT$ of energy. But doubling the temperature doubles the size of a chunk of energy that can be easily moved. This leads to the conclusion that if two oscillators have frequencies $f$ and $2f$, the one at $2f$ must emit energy in quanta just twice the size of those emitted by the one at $f$. The size of the energy quantum therefore depends linearly on the frequency,

$$E = hf$$

where $h$ is Planck's constant.

The way this evolved historically was that at first the higher frequency part of the black body radiation curve was mapped out by experimentalists in Berlin, and Wien showed that it was well approximated by the formula

$$R_W(f)\,df \,\alpha\, f^3 e^{-Af}\,df$$

with $A$ a constant. This was very reminiscent of the Maxwell distribution for gas molecules, but with $f^3$ in place of $v^2$. It soon became evident, though, that the formula was not accurate at low frequencies, where in fact it appeared that the distribution went as $f^2$, as discussed above. Planck then wrote down a formula having the right behaviour for both high and low frequencies:

$$R_T(f)\,df = \frac{8\pi V h f^3 df / c^3}{e^{hf/kT} - 1}$$

He did not at first have any theoretical justification for this formula, but it was a very accurate fit to some very precise experiments.

It is easy to see how Wien's Displacement Law follows from this formula the maximum radiation per unit frequency range is at the frequency $f$ for which the function $f^3/(e^{hf/kT}-1)$ is a maximum. Solving numerically gives $hf_{max}=2.82kT$. It is perhaps worth mentioning that this is the frequency range at which the radiation *per unit frequency range* is a maximum.Wien's law is written in terms of wavelengths, and gives the *wavelength* at which the *radiation per unit wavelength range* is maximum. Because the wavelength and frequency are not linearly related, but in fact inversely related, these two maxima are not at the same frequency

## JUSTIFYING PLANCK'S FORMULA

Factoring out the number of modes of oscillation in the frequency range df, we see that Planck's formula tells us that the average energy per mode is

$$\frac{hf}{e^{hf/kT}-1}$$

For low frequencies, $hf \ll kT$, this correctly gives $kT$ per mode. The formula follows if we assume the oscillator can only have a whole number of quanta of energy, that is to say, its energy must be one of: 0, $hf$, $2hf$, $3hf$, ... If we further assume that the relative probability of it having energy $E$ is $e^{-E/kT}$, then its relative probabilities of having energy 0, $hf$, $2hf$, ... are in the ratio 1: $e^{-hf/kT}$: $e^{-2hf/kT}$, etc. The actual probabilities are given by dividing these relative probabilities by the sum of all of them They clearly are the terms of a geometric series, so their sum is just $1/(1 - e^{-hf/kT})$. So, to find the average energy in the oscillator, we take the possible energies 0, $hf$, $2hf$, $3hf$, ... and weight each of them with their probability of occurring, that is, we must find $0.1 + hf.\ e^{-hf/kT} + 2hf.\ e^{-2hf/kT} + \ldots$, and divide the sum by $1/(1 - e^{-hf/kT})$. This is left as an exercise for the reader. Planck himself gave a more complete and elegant discussion, applying Boltzmann's ideas about entropy to simple harmonic oscillators having only certain energies allowed. In fact, he effectively derived the Boltzmann distribution for this system.

# Chapter 15

# Applications

As in other scientific fields, nuclear scientists generally work with a great interest and excitement for the science. Understanding the building blocks of nature and the physical laws that govern them is the ultimate goal. On the other hand, members of the public are the taxpayers who finance most research in basic science and, understandably, often want to see something more concrete emerge from their investments, some further benefit to society. In the century since Rutherford discovered the nucleus, numerous applications take advantage of one or more of the following:

1. The properties of nuclei,
2. Measurement techniques developed in nuclear physics,
3. Particle accelerators, and
4. Other tools of nuclear science such as detectors.

These applications benefit disciplines as diverse as medicine, biology, art, archaeology, energy, materials science, space exploration, and the environment. Only, a sample of them will be described in this chapter.

The following examples are illustrative of the wide variety of applications of nuclear science. While some of these applications are very specific and have little effect on our daily lives, many of these applications are all pervasive. We probably use some electricity generated by nuclear reactors, protect our homes with smoke alarms, and have all been affected by the geopolitics of nuclear weapons.

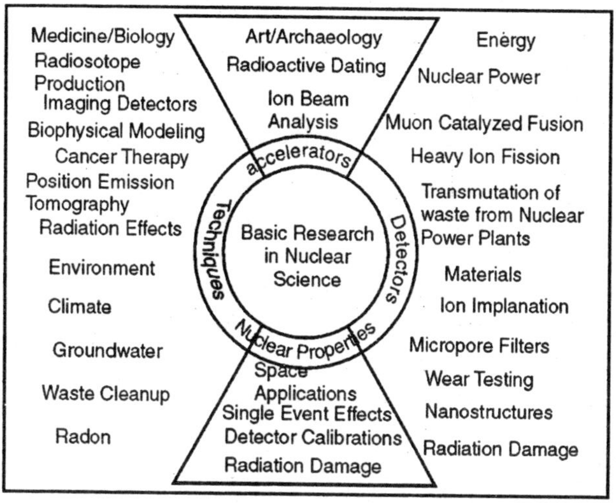


## Smoke Detectors

Most common smoke detectors contain a small amount of $^{241}$Am, a radioactive isotope. $^{241}$Am is produced and recovered from nuclear reactors. Alpha particles emitted by the decays of $^{241}$Am ionize the air (split the air molecules into electrons and positive ions) and generate a small current of electricity that is measured by a current-sensitive circuit. When smoke enters the detector, ions become attached to the smoke particles, which causes a decrease in the detector current. When this happens, an alarm sounds. These detectors provide warning for people to leave burning homes safely. Many lives have been saved by the their use.

Because the distance alpha particles travel in air is so short, there is no risk of being exposed to radiation by having a smoke detector in the house. Since ionization-type smoke detectors contain radioactive materials, they should be recycled or disposed of as radioactive waste. It is important to follow the instructions that come with the smoke alarm when they need to be discarded.

## Space Applications: Radiation-Induced Effects

Radiation-induced spacecraft anomalies have been known since the Explorer I launch on January 31, 1958, when a Geiger counter put aboard by J.A. Van Allen suddenly stopped counting. It turned out that the counter was in fact saturated by an extremely high count rate. This led to the discovery of the Van Allen belts.

The inner belt, beginning at about 1,000 km above the surface of the Earth, contains primarily protons with energies between 10-100 MeV. The offset between Earth's geographical and magnetic axes causes an asymmetry in the radiation belt above the Atlantic Ocean off the Brazilian coast, allowing the inner belt to reach a minimum altitude of 250 km. This "South Atlantic Anomaly" is important because it occupies a region through which low-orbiting satellites spend as much as 30percent of their time. During a solar flare, which can happen anytime, the number of protons suddenly increases by more than a million.

The first spacecraft loss due to total radiation dose effects occurred unexpectedly in 1962. A satellite, Telstar, was launched just one day after a high altitude nuclear weapons test. This weapons test produced a large number of beta particles, which caused a new and very intense radiation belt that lasted until the early 1970s. Telstar and six other satellites were lost within a seven-month period after this weapons test. Telstar was well-studied and the loss was traced directly to breakdown of diodes in the command decoder due to the total radiation dose.

Radiation effects studies done at accelerators measure either total-dose effects, displacement damage, or single event effects (SEE). A SEE occurs when one ion passing through a semiconductor causes enough damage to upset the circuit in some way. In 1978, the first SEEs were observed at ground level when Intel Corporation discovered that anomalous upsets occurring in dynamic random access computer memories were being caused by alpha particles emitted from trace amounts of thorium and uranium in the materials from which the device's packages were made. It was quickly found that massive ions, protons, and neutrons could all induce SEEs.

SEEs can result in the flip of a memory bit as in the case of upsets with little other effect on the circuit, or they can be more catastrophic, causing the chip to either latchup and have to be powered down to reset the circuit, or it might burn out altogether or rupture a gate.

## Biology Studies

Radiation biologists study the effect of radiation on living tissue. This field, which has direct analogies to the SEE studies, looks at the damage sustained by cells in the cosmic ray environment of space, a single event effect. These studies are necessary before we send astronauts to Mars.

## Materials Studies

In 1959, it was discovered that mica, that was exposed to fission fragments, showed "nuclear tracks" when viewed under an ordinary microscope. This discovery led to a new class of nuclear physics detectors as well as a wide range of practical applications in earth science, oceanography, biology, medicine, archaeology, and space science.

By exposing thin sheets of plastic or other material to massive ions, tracks can be made through the material and then etched to make very fine "micropore" filters (30 nm to 8 μm hole size). These have been made at reactors since about 1970 and are used primarily in the biomedical field, with other applications in geophysics, meteorology, and brewing.

Another use for these sheets of plastic is to measure radon levels. By counting the number of ion tracks left in the plastic, one can measure the radon concentration.

## Nuclear Medicine: Radioisotopes for Diagnosis and Treatment

One major use of radioisotopes is in nuclear medicine. Of the 30 million people who are hospitalized each year in the United States, 1/3 are treated with nuclear medicine. More than 10 million nuclear-medicine procedures are performed on patients and more than 100 million nuclear-medicine tests are performed each year in the United States alone. A comparable number of such procedures are performed in the rest of the world.

There are nearly one hundred radioisotopes whose beta and/or gamma radiation is used in diagnosis, therapy, or investigations in nuclear medicine. The most used radioisotopes were discovered before World War II using the early cyclotrons of Ernest Lawrence, with the initial applications to medicine being developed by his brother John Lawrence. Some of the most well known radioisotopes, discovered by Glenn Seaborg and his coworkers, are $^{131}I$ (discovered in 1938), $^{60}Co$ (1937), $^{99m}Tc$ (1938), and $^{137}Cs$ (1941). By 1970, 90 percent of the 8 million administrations per year of radioisotopes in the United States utilized either $^{131}I$, $^{60}Co$, or $^{99m}Tc$. Today, $^{99m}Tc$, with a half-life of 6 hours, is the workhorse of nuclear medicine. It accounts for more than 10 million diagnostic procedures a year in the United States. It is used for brain, bone, liver, spleen, kidney, lung and thyroid imaging as well as for blood-flow studies.

$^{131}I$, with a half-life of 8 days, is used to diagnose and treat thyroid disorders. Seaborg's mother was one of the first to benefit from the use of this radioisotope that her son had discovered. Fatally ill from hyperthyroidism, (a related condition from which her sister died), diagnosis and treatment with $^{131}I$ led to her complete recovery and a long life. Former President George Bush and First Lady Barbara Bush are some notable people who were successfully treated for Graves' disease, a thyroid disease, with $^{131}I$. Radioactive iodine treatment is so successful that it has virtually replaced thyroid surgery.

A very effective role for radioisotopes in nuclear medicine is the use of short-lived positron emitters such as $^{11}C$, $^{13}N$, $^{15}O$, or $^{18}F$ in a process known as Positron Emission Tomography (PET). Incorporated in chemical compounds that selectively migrate to specific organs in the body, diagnosis is effected by detecting annihilation gamma rays–two gamma rays of identical energy emitted when a positron and an electron annihilate each other. These gamma rays have the very useful property that they are emitted in exactly opposite directions. When both are detected, a computer system may be used to reconstruct where the annihilation occurred. By attaching a positron emitter to a protein or a glucose molecule, and allowing the body to metabolize it, we can study the functional aspect of an organ such as the human brain. The PET image shows where the glucose has been absorbed.

PET imaging becomes even more valuable when we can observe the functional image compared to the anatomical image. Magnetic Resonance Imaging (MRI)–originally known as Nuclear Magnetic Resonance Imaging–can provide very detailed images of the anatomy. These techniques provide researchers a better understanding of what is healthy tissue versus what is diseased.

**Nuclear Medicine: Cancer Therapy**

The radioisotope $^{60}$Co emits gamma rays that are used to destroy cancer cells. Hundreds of thousands of Americans who suffer from cancer have been treated in this way. Every year millions of cubic meters of medical products and equipment are sterilized by irradiation worldwide. $^{137}$Cs has found substantial applications as a gamma-ray source in medical therapy, similar in its use to that of $^{60}$Co.

Cancer treatment with beams of massive ions directly from an accelerator has gained increasing utilization in the last decade. Unlike gamma rays, which distribute their energy equally in healthy as well as cancerous cells, massive particles such as protons or alpha particles will deposit the bulk of their energy just before they stop. If the energy is well-chosen, most of the energy will be dumped into the tumor and not into the surrounding healthy tissue. Using three dimensional water degrader columns, the shape of the tumor can be mapped out and selectively irradiated. Dedicated accelerators are now being built to continue this work at medical centres. In the United States, the Loma Linda Medical Centre in California is now operating a proton synchrotron for therapy, and a second facility is being built at Massachusetts General in Boston.

Boron Neutron Capture Therapy (BNCT) is under development for the treatment of glioblastoma multiforma, a brain cancer which afflicts some 12,500 people a year in the US alone and is almost impossible to treat by currently available means. In BNCT, boron is synthesized into compounds that are selectively taken up by cancerous cells in the brain and not by healthy ones. In subsequent irradiation by low-energy neutrons (from a reactor or accelerator), the following neutron-capture reaction occurs:

$^{10}B + n$ Æ $^{11}B$ Æ $^{7}Li + {}^{4}He$.

The recoiling lithium and helium nuclei have short ranges and large energy losses. These particles destroy the cancerous cells and spare the surrounding healthy tissue.

**Radioactive Dating**

The technique of comparing the abundance ratio of a radioactive isotope to a reference isotope to determine the age of a material is called radioactive dating. Many isotopes have been studied, probing a wide range of time scales.

The isotope $^{14}C$, a radioactive form of carbon, is produced in the upper atmosphere by neutrons striking $^{14}N$ nuclei. The neutron is captured by the $^{14}N$ nucleus and knocks out a proton. Thus, we have a different element, $^{14}C$. The isotope, 14C, is transported as $^{14}CO_2$, absorbed by plants, and eaten by animals. If we were to measure the ratio of $^{14}C$ to $^{12}C$ today, we would find a value of about one $^{14}C$ atom for each one-trillion $^{12}C$ atoms. This ratio is the same for all living things–the same for humans as for trees or algae.

Once living things die, they no longer can exchange carbon with the environment. The isotope $^{14}C$ is radioactive, and beta-decays with a half-life of 5,730 years. This means that in 5,730 years, only half of the $^{14}C$ will remain, and after 11,460 years, only one quarter of the $^{14}C$ remains. Thus, the ratio of $^{14}C$ to $^{12}C$ will change from one in one-trillion at the time of death to one in two trillion 5,730 years later and one in four-trillion 11,460 years later. Very accurate measurements of the amount of $^{14}C$ remaining, either by observing the beta decay of $^{14}C$ or by accelerator mass spectroscopy (using a particle accelerator to separate $^{12}C$ from $^{14}C$ and counting the amount of each) allows one to date the death of the once-living things.

Perhaps you have heard of Ice Man, a man living in the Alps who died and was entombed in glacial ice until recently when the ice moved and melted. The man's body was recovered and pieces of tissue were studied for their $^{14}C$ content by accelerator mass spectroscopy. The best estimate from this dating technique says the man lived between 3350 and 3300 BC.

The boat of a pharaoh was discovered in a sealed crypt and reassembled in a museum near the pyramids. Its wood was dated using $^{14}C$ to be about 4,500 years old.

**Fig.** The Pharaoh's Funerary Boat.

Other methods of dating are used for non-living things. $^{40}K$ decays with a half-life of $1.3 \times 10^9$ years to $^{40}Ar$ which can be trapped in rocks. A potassium-argon method of dating, developed in 1966, measures the amount of $^{40}Ar$ arising from the $^{40}K$ decay and is compared to the amount of $^{40}K$ remaining in the rock. From the ratio, the time since the formation of the rock can be calculated.

The age of our galaxy and earth also can be estimated using radioactive dating. Using the decays of uranium and thorium, our galaxy has been found to be between 10 and 20 billion years old and the earth has been found to be 4.6 billion years old. The Universe must be older than our galaxy. Within experimental error, this estimate agrees with the 15 billion year estimate of the age of the Universe.

### Neutron Activation Analysis

There are at least 50 elements occurring in nature that have radioactive isotopes with one neutron more than their stable isotopes. This means that the radioactive species can be made by neutron bombardment. This procedure is typically done in a nuclear reactor, although other neutron sources can sometimes be used. The stable nucleus absorbs one neutron and becomes a radioactive nucleus. By detecting the decay of these nuclei, which can be done with great sensitivity, one can measure the concentration of the stable element of interest in the sample.

A simple example of neutron activation analysis involves the measurement of iridium in soils. This is the measurement that led to the theory that the extinction of the dinosaurs, 65 million years ago, was caused by the impact of an asteroid or comet somewhere on Earth. An impact would produce so much debris in the air that earth would receive a large reduction in sunlight and thus fewer plants would grow. With less food available, there would be a devastating reduction in number of animals. This reduction was so severe that most species, including the dinosaurs, became extinct.

In 1979, a group of scientists reported that neutron activation analysis had shown unusual amounts of the element iridium in Italian Cretaceous-Tertiary boundary sediments. Since iridium is a metal which has very low abundance on earth, they attributed the excess iridium to a impact of a 10-kilometre diameter asteroid. This work, which melded the disciplines of physics (Nobel Laureate Luis Alvarez), geology (his son, Walter Alvarez) and nuclear chemistry (Isadore Perlman, Frank Asaro and Helen Michel), galvanized the scientific world because the concepts presented could be tested by many diverse techniques.

Since the initial report, anomalous amounts of iridium have been found at the Cretaceous-Tertiary boundary in over 100 sites worldwide. Many experiments have confirmed its impact origin.

## Industrial Applications

The applications of radioisotopes in industry are numerous. Many types of thickness gauges exploit the fact that gamma rays are attenuated when they pass through material. By measuring the number of gamma rays, the thickness can be determined. This process is used in common industrial applications such as:

1. The automobile industry–to test steel quality in the manufacture of cars and to obtain the proper thickness of tin and aluminum
2. The aircraft industry–to check for flaws in jet engines
3. Construction–to gauge the density of road surfaces and subsurfaces
4. Pipeline companies–to test the strength of welds
5. Oil, gas, and mining companies–to map the contours of test wells and mine bores, and
6. Cable manufacturers–to check ski lift cables for cracks.

The isotope $^{241}$Am is used in many smoke detectors for homes and businesses (as mentioned previously), in thickness gauges designed to measure and control metal foil thickness during manufacturing processes, to measure levels of toxic lead in dried paint samples, and to help determine where oil wells should be drilled.

The isotope $^{252}$Cf (a neutron emitter) is used for neutron activation analysis, to inspect airline luggage for hidden explosives, to gauge the moisture content of soil and other materials, in bore hole logging in geology, and in human cervix-cancer therapy.

In addition, there are manifold uses in agriculture. In plant research, radiation is used to develop new plant types to speed up the process of developing superior agricultural products. Insect control is another important application; pest populations are drastically reduced and, in some cases, eliminated by exposing male insects to sterilizing doses of radiation. Fertilizer consumption has been reduced through research with radioactive tracers. Radiation pellets are used in grain elevators to kill insects and rodents. Irradiation prolongs the shelf-life of foods by destroying bacteria, viruses, and molds.

The useful application of radioisotopes extends to the arts and humanities. Neutron activation analysis is extremely useful in identifying the chemical elements present in coins, pottery, and other artifacts from the past. A tiny unnoticeable fleck of paint from an art treasure or a microscopic grain of pottery suffices to reveal its chemical makeup. Thus the works of famous painters can be "fingerprinted" so as to detect the work of forgers.

Neutron scattering has proved to be a valuable tool for studying the molecular structure and motion of molecules of interest to manufacturing and life processes. Accelerators and reactors produce low-speed neutrons with wavelength appropriate to "see" structures of the size of magnetic microstructures and DNA molecules. Neutrons can penetrate deeply into bulk materials and use their magnetic moment or strong interaction forces to preferentially scatter from magnetic domains or hydrogen atoms in long chain nucleosomes. Neutrons are also used in materials surface and interface studies taking advantage of their reflectivity properties. Intense sources of neutrons include: the IPNS at Argonne National Laboratory in Illinois and LANSCE at Los Alamos National Laboratory in New Mexico.

### Removal of Land Mines

Wars have been fought around the globe, and land mines were used extensively by the battling parties. There may be up to 100 million abandoned mines just waiting to maim and kill unsuspecting civilians. Techniques similar to airport weapons detectors have been considered for finding and neutralizing these mines without loss of life.

A promising technology involves nuclear quadrupole resonance. Nitrogen, a common component in explosives, has a nucleus with an ellipsoidal shape. Depending on what kind of crystalline structure the nitrogen nuclei find themselves in, their non-sphericity produces a unique set of very narrowly spaced energy levels that is characteristic of the crystalline solid itself. An explosive compound can therefore be identified by the subtle effect of its constituent nitrogen atoms. This technique could also differentiate explosives from scrap metals in the ground. More details may be found in the reference at the end of this chapter.

### Radioisotope Power Generation

Long lived power sources are needed for equipment that is too remote or inaccessible for replacement. By choosing a radioactive element with a long half life, we can create a long lived power source. The appropriate element should:

1. Produce weakly penetrating radiation that can be easily shielded
2. A specific power of at least 0.2 kW/kg
3. Have good corrosion resistance
4. Be insoluble in water
5. Be made of reasonably available material.

Among the transuranium elements, oxides of the alpha-emitting nuclides $^{238}Pu$ (t1/ 2 = 87.7 years) and $^{244}Cm$ ($t_{1/2}$ = 18.1 years) are useful fuels. A few grams to kilograms of such nuclides, in appropriately shielded containers, provide intense sources of heat with power levels up to hundreds of watts, since the alpha particles are stopped very easily and their decay energy converted into thermal energy. Using thermoelectric devices without moving parts, it is possible to convert the resultant heat flow into usable electricity.

Such power sources are small, lightweight, and rugged. One of their uses is in the SNAP (Space Nuclear Auxiliary Power) units used to power satellites, or more importantly, to power remote sensing instrument packages. SNAP sources (fueled by $^{238}Pu$) served as the power sources for instrument packages on the five Apollo missions, the Viking unmanned Mars lander, and the Pioneer, Voyager and Cassini probes to Jupiter, Saturn, Uranus, Neptune, Pluto, and beyond.

In addition to space applications, radionuclide power sources have been used as terrestrial sources of energy wherever compact and long-lived sources, not requiring maintenance, are needed. In the early 1970s, $^{238}Pu$ batteries were used as the power sources for cardiac pacemakers. Over 3,500 units were implanted and most remain functioning. Due to a lower cost and easier construction, lithium batteries have largely supplanted the plutonium.

**Nuclear Weapons**

Without a doubt, the development of nuclear weapons is one application of nuclear science that has had a significant global influence. Following the observation of fission products of uranium by Hahn and Strassmann in 1938, a uranium fission weapon became possible in the eyes of a number of nuclear scientists. It was Albert Einstein who signed a letter to Franklin Roosevelt, President of the United States, and alerted him to the potential development of a nuclear weapon. Thus began the Manhattan Project, resulting in production of the first nuclear weapons.

In 1945, a bomb using the fission of $^{235}U$ was dropped on Hiroshima, while a bomb using the fission of $^{239}Pu$ was dropped on Nagasaki. The awesome destruction that was produced by two nuclear devices changed the face of warfare forever. Not only did these two explosions end the most destructive war in history, but the possessor of these weapons of mass destruction seemed invincible to any adversary.

The legacy of nuclear weapons remains in the soil around us, where we can still measure minute amounts of fallout from the atmospheric nuclear tests of the 1950s and 1960s. The weapons need constant maintenance because radiation damage affects the materials and triggering devices that must act together to make the weapon work. Furthermore, the isotope tritium, which is used in weapons, naturally decays. Storage and disposal of the weapons is technically difficult but probably a manageable problem. However, it is a daunting political and social problem. It is a challenge to maintain these weapons, to significantly reduce their numbers, to clean up the waste, and to insure that these weapons never will be used again.

Nuclear weapons are with us to this day and could be with us in the future. Even if all conflicts among nations end, some nuclear weapons might be retained. Earth is vulnerable to impacts from comets and asteroids. Some scientists have proposed that nuclear weapons could possibly be used to deflect them from hitting our planet.

## Agricultural Applications

Radioisotopes can be used to help understand chemical and biological processes in plants. This is true for two reasons: 1)radioisotopes are chemically identical with other isotopes of the same element and will be substituted in chemical reactions and 2)radioactive forms of the element can be easily detected with a Geiger counter or other such device.

Example:

A solution of phosphate, containing radioactive phosphorus-32, is injected into the root system of a plant. Since phosphorus-32 behaves indentically to that of phosphorus-31, the more common and non-radioactive form of the element, it is used by the plant in the same way. A Geiger counter is then used to detect the movement of the radioactive phosphorus-32 throughout the plant. This information helps scientists understand the detailed mechanism of how plants utilized phosphorus to grow and reproduce.

## Food Irradiation

Food irradiation is a method of treating food in order to make it safer to eat and have a longer shelf life. This process is not very different from other treatments such as pesticide application, canning, freezing and drying. The end result is that the growth of disease-causing microorganismns or those that cause spoilage are slowed or are eliminated altogether. This makes food safer and also keeps it fresh longer.

Food irradiated by exposing it to the gamma rays of a radioisotope — one that is widely used is cobalt-60. The energy from the gamma ray passing through the food is enough to destroy many disease-causing bacteria as well as those that cause food to spoil, but is not strong enough to change the quality, flavor or texture of the food. It is important to keep in mind that the food never comes in contact with the radioisotope and is never at risk of becoming radioactive.

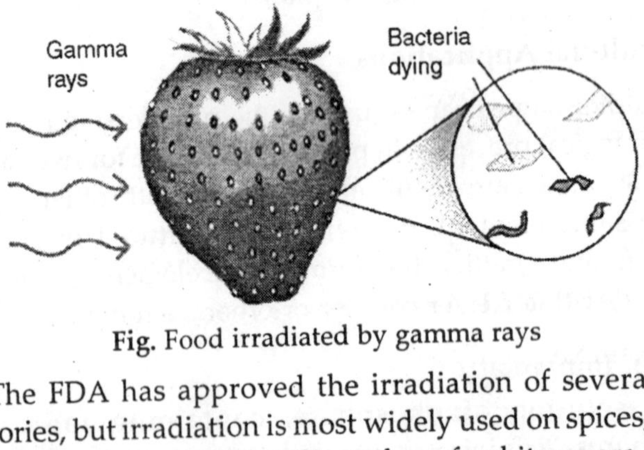


Fig. Food irradiated by gamma rays

The FDA has approved the irradiation of several food catagories, but irradiation is most widely used on spices, herbs and dehydrated vegetables. Since these food items are grown in or on the ground it is clear to see that they are at risk for being exposed to naturally occuring pests such as bacterias, molds, fungi, insects, and rodents. It is impossible to harvest and package these items without having some contamination from these naturally occuring pathogens. Irradiation of this material can help to ensure that the final product you purchase is pest free.

Some meats are irradiated. Pork, for example, is irradiated to control the trichina parasite that resides in the muscle tissue of some pigs. Poultry is irradiated to eliminate the chance of foodborne illness due to bacterial contamination.

Irradiation of certain foods also have additional benefits. Since the energy passing through the food can disrupt cellular processes (this is the mechanism for destroying microorganisms) it also can halt the cellular processes that lead to the sprouting or ripening of foods. Potatoes and onions are irradiated to retard their sprouting. Fruits and vegetables are irradiated to slow down the ripening process. In this way, delicate fruits won't reach their peak ripeness before they arrive at the supermarket.

Food irradiation sounds like a wonderful use of nuclear chemistry principles. Although this process is routinely used in Europe, Canada, and Mexico, the United States has been a little more hesitant to adopt food irradiation. This is due mainly to the public's perceived fear and limited understanding of nuclear science. For example, although the FDA has approved the irradiation of poultry, the industry hesitates to adopt the process because they are afraid of a negative response from consumers. With recent public education, however; many people are learning to appreciate and value it's usefullness.

## INDUSTRIAL APPLICATIONS

### Ion Implantation in Semiconductor

Ion implantation is a process in which ions of a material can be implanted to another solid, thereby changing the physical properties of the solid. Semiconductors like silicon can alter the conductivity by introducing a small quantity of dopant atoms like boran, arsenic phosphorus. It is important to accurately control the number of implanted dopants and to place them at a desired depth. The ion implantation equipment typically consists of an ion source, an accelerator and a target chamber. The ions of the desired element are produced and these are accerelerated in the electric field to a high energy. A magnet selects the desired beam and directs it towards the target, which is the material to be implanted.

## Material Analysis

### *Isotope Dating*

Radioactive isotope dating is a sophisticated method of determining the age of a material. This is done by making a comparison between the abundance ratio of a radioactive isotope to a reference isotope. The radioactive isotope C-14 has a half life period of 5,730 years. The half-life of a radioactive isotope is the time it takes for half the isotopes in the sample to decay away. So, for a sample of C-14, after 5,730 years only half of the original number of 14C will remain, and after 11,460 years, only one quarter of the 14C remains. Very accurate measurements of the amount of carbon-14 in a material can be done either by observing the beta decay of 14C or by accelerator mass spectroscopy (using a particle accelerator).

### *Accelerator Mass Spectroscopy*

The basic principle in accelerator mass spectrometry is to separate various atoms according to their charge-to-mass ratio. An ion source bombards the sample to be dated with a cesium beam that negatively ionizes and releases the carbon atoms from the sample. Impurities are also present. A magnet and slit arrangement gets rid of some of the impurities before the negative beam is injected into an accelerator with a field of millions of volts. The accelerated ions hit a thin carbon foil that strips off electrons, and in so doing, destroys all molecular species. The carbon atoms, now positively charged ions, continue to be accelerated. On exiting the accelerator, the ions are passed through a mass spectrometer and various filters that separate the particles according to their mass. Finally, the ions slow down to a stop in a gas ionization detector according to their charge. The identity of individual ions can be determined from how the ions slow down. For example, carbon-14 slows down more slowly than nitrogen-14, so those ions of the same mass can be distinguished from one another. Once the charges are determined, the detector can tell to which element each ion belongs and counts the desired isotope as a ratio of a more abundant isotope-carbon-14.

### *Sterilization*

Particle accelerators are use for medical sterilization, food preservation, purifying drinking water and removing noxious substances. There is lot of food born illness erupting which could be mitigated through the sterilization of food using accelerators; but this method is still not widely accepted by the general public. Irradiation is the most effective method of eliminating the deadly bacteria in raw meat. It can also significantly reduce levels of other pathogens, including Listeria, Salmonella, Shigella, Campylobacter and others. While there is no single technique to cure all food problems, irradiation is accepted as both safe and effective. Linear accelerators can be used for irradiation. There are also some sterilization processes which accelerators can perform. An accelerator is an alternative to the autoclave to sterilize surgical instruments and laundry. Also, the preservation of food stuff for animals and production of fertilizer for crops is commonly done, as these are perceived to be sufficiently remote from human ingestion, so that the public is not afraid of a radiation hazard.

### *Microstructure Lithography*

X-rays from light sources can be focused to such a fine point that they can cut out tiny defects such as machine components from silicon or plastic with incredible precision on a scale of thousandths of a millimeter. They can also be used to etch patterns in microchips

### *Explosives*

Particle accelerators produce charged particles and accelerate them into a high energy beam; particle beams may be used for many different applications. In the 1950s scientists who were doing research in the material sciences found that particle accelerators can be used in producing heavy metals such as Uranium-233 and Plutonium-239. While studying cosmic rays it was discovered that high energy particles can be used to produce neutrons. This enabled the use particle accelerators for the production of fission material used in explosives. Fission material such as uranium-233 is produced from fertile material thorium-232; when an atom of thorium-232 absorbs a neutron, Uranium-233 is produced. Similarly plutonium-239 can be produced from Uranium-238 by absorbing a neutron.

### *Transmutation of Nuclear Waste*

Apart from developing fission material for explosives, particle accelerators can also be used for treating nuclear waste. Nuclear waste from nuclear reactors is stored in storage pools, but nowadays the storage pools are filled completely and the radioactive materials in the waste have half-life period of million years. Since there is no way to dispose of the nuclear waste without causing a threat, a new concept called ADTT (Accelerator Driven Transmutation Technology) has been developed. It can reduce the half-life period of the radioactive waste to a few centuries which would be helpful in storing the waste. The conceptual basis is to use neutrons to transmute the particles of radioactive elements such as Plutonium and Neptunium to products with a lower half-life. Accelerator based conversion can also be used to destroy plutonium from weapons and reactor fuel sources by producing products with lower half-life. This process can burn up to 98% of plutonium-239 and 90% of plutonium isotopes from explosive material.